Analysis Techniques
for Information Security

Synthesis Lectures on Information Security, Privacy, and Trust

Editor
Ravi Sandhu, *University of Texas, San Antonio*

Synthesis Lectures on Information Security, Privacy, and Trust is composed of 50 to 100-page publications on topics pertaining to all aspects of the theory and practice of Information Security, Privacy, and Trust. The scope will largely follow the purview of premier computer security research journals such as ACM Transactions on Information and System Security, IEEE Transactions on Dependable and Secure Computing, and Journal of Cryptology, and premier research conferences, such as ACM CCS, ACM SACMAT, ACM AsiaCCS, IEEE Security and Privacy, IEEE Computer Security Foundations, ACSAC, ESORICS, Crypto, EuroCrypt and AsiaCrypt. In addition to the research topics typically covered in such journals and conferences, the series also solicits lectures on legal, policy, social, business and economic issues addressed to a technical audience of scientists and engineers. Lectures on significant industry developments by leading practitioners are also solicited.

Analysis Techniques for Information Security
Anupam Datta, Somesh Jha, Ninghui Li, David Melski, and Thomas Reps
2010

Operating System Security
Trent Jaeger
2008

Analysis Techniques for Information Security

Anupam Datta, Somesh Jha, Ninghui Li, David Melski, and Thomas Reps

www.morganclaypool.com

ISBN: 9781598296297   paperback
ISBN: 9781598296303   ebook

DOI 10.2200/S00260ED1V01Y201003SPT002

A Publication in the Morgan & Claypool Publishers series
SYNTHESIS LECTURES ON INFORMATION SECURITY, PRIVACY, AND TRUST

Lecture #2
Series Editor: Ravi Sandhu, *University of Texas, San Antonio*
Series ISSN
Synthesis Lectures on Information Security, Privacy, and Trust
Print 1945-9742   Electronic 1945-9750

Analysis Techniques for Information Security

Anupam Datta
Carnegie Mellon University

Somesh Jha
University of Wisconsin

Ninghui Li
Purdue University

David Melski
GrammaTech, Inc.

Thomas Reps
University of Wisconsin and GrammaTech, Inc.

SYNTHESIS LECTURES ON INFORMATION SECURITY, PRIVACY, AND TRUST #2

MORGAN & CLAYPOOL PUBLISHERS

ABSTRACT

Increasingly our critical infrastructures are reliant on computers. We see examples of such infrastructures in several domains, including medical, power, telecommunications, and finance. Although automation has advantages, increased reliance on computers exposes our critical infrastructures to a wider variety and higher likelihood of accidental failures and malicious attacks. Disruption of services caused by such undesired events can have catastrophic effects, such as disruption of essential services and huge financial losses. The increased reliance of critical services on our cyberinfrastructure and the dire consequences of security breaches have highlighted the importance of information security. Authorization, security protocols, and software security are three central areas in security in which there have been significant advances in developing systematic foundations and analysis methods that work for practical systems. This book provides an introduction to this work, covering representative approaches, illustrated by examples, and providing pointers to additional work in the area.

KEYWORDS

information security, static analysis, software security, security policies, protocol verification

Contents

Acknowledgments

The material in Section 4.2 originally appeared in [140], and the material in 4.4 originally appeared in [162]. The material in §2.2 originally appeared in [215]. The work on Protocol Composition Logic described in the chapter on security protocol analysis is based on A. Datta's joint work with A. Derek, J. C. Mitchell, D. Pavlovic, and A. Roy.

We thank Denis Gopan for permission to use background material from his Ph.D. thesis [118, Chapter 2] in App. A.

Anupam Dutta was supported in part by the U.S. Army Research Office contract on Perpetually Available and Secure Information Systems (DAAD19-02-1-0389) to Carnegie Mellon CyLab and the NSF Science and Technology Center TRUST.

Somesh Jha was supported in part by NSF under grants CNS-0904831, CCF-0524051, and CNS-0448476

Ninghui Li was supported in part by NSF under grants CNS-0448204 and CCR-0325951.

Thomas Reps was supported in part by NSF under grants CCF-0540955 and CCF-0524051, by HSARPA under AFRL contract FA8750-05-C-0179, and by ONR under grant N00014-01-1-0796.

Anupam Datta, Somesh Jha, Ninghui Li, David Melski, and Thomas Reps
February 2010

CHAPTER 1

Introduction

Increasingly, our critical infrastructures are becoming heavily dependent on computers. We see examples of such infrastructures in several domains, including medical, power, telecommunications, and finance. Although automation provides society with the advantages of efficient communication and information sharing, the pervasive, continuous use of computers exposes our critical infrastructures to a wider variety and higher likelihood of accidental failures and malicious attacks. Disruption of services caused by such undesired events can have catastrophic effects, including loss of human life, disruption of essential services, and huge financial losses. For example, the outbreak of the CodeRed virus infected more than 359, 000 hosts, resulting in financial losses of approximately 2.6 billion dollars [195]. The increased reliance of critical services on our cyberinfrastructure and the dire consequences of security breaches have highlighted the importance of information security.

Authorization, *security protocols*, and *software security* are three central areas in security in which there have been significant advances in developing systematic foundations and analysis methods that work for practical systems. This book provides an introduction to this work, covering representative approaches, illustrated by examples, and providing pointers to additional work in the area. We describe the kinds of questions that drive research in these areas and the general methods for security analysis.

The rest of the book is organized as follows. Chapter 2 presents two general program analysis methods—*static analysis* and *logic programming* (with Datalog)— that have been applied to a broad range of problems in information security. As a representative example of systematic approaches to software security, Chapter 3 describes a static-analysis method for detecting buffer overruns (a serious security vulnerability) in C programs. Chapter 4 illustrates the use of static analysis and Datalog for representing and analyzing authorization policies. Finally, Chapter 5 discusses the general methodology of security-protocol analysis and illustrates it using a specialized protocol logic. We elaborate below on the contents of each of these chapters.

Program Analysis Methods (Chapter 2). The first part of Chapter 2 discusses static analysis, which is a technique to obtain information about the possible reachable states of a program without actually executing the program. Static-analysis methods are well-suited in the context of information security because they reason about all possible executions (including those involving inputs provided by a malicious attacker) about a program. Static analysis has been applied to several problems in information security, such as malware detection [66], generating signatures [51; 83], and discovering vulnerabilities [57; 63]. Specifically, we describe interprocedural dataflow analysis, and discuss a formalism called *weighted pushdown systems (WPDSs)*, which can be used to formulate a variety of interprocedural dataflow-analysis problems that are relevant in addressing security problems.

The second part of the chapter is devoted to Datalog, a logic programming language. In the context of information security, Datalog has been used for knowledge representation and analysis. For example, Datalog is the basis of several languages for describing security policies [92; 143; 158], and it has been used for vulnerability analysis of networks and systems [62; 206].

Detecting Buffer Overruns (Chapter 3).   Buffer overruns are one of the most exploited classes of security vulnerabilities. Malicious attacks that exploit buffer overruns can have serious consequences, such as a remote user acquiring root privileges on a host. This chapter describes a static-analysis technique for detecting buffer overruns in C programs. We describe the design and implementation of a tool that statically analyzes C source code to detect buffer-overrun vulnerabilities. One of our principal design goals was to make the tool scale to large real-world applications. We used the tool to audit several popular and commercially used packages. The tool identified 14 previously unknown buffer overruns in `wu-ftpd-2.6.2` (Section 3.5.1.1) in addition to several known vulnerabilities in other applications.

Analyzing Security Policies (Chapter 4).   Systems with shared computing resources use access-control mechanisms for protection. The main issues in access control are *authentication*, *authorization*, and *enforcement*. Identification of principals is handled by authentication. Authorization addresses the question, "Should a request *r* by a specific principal *A* be allowed?" Enforcement addresses the problem of implementing the authorization during an execution. We focus on authorization. Given a security policy for authorization, *safety analysis* decides whether rights can be leaked to unauthorized principals in future states. In this chapter, we formally define the security analysis problem and survey the complexity of these problems for classic access control mechanisms. We also discuss safety analysis for *Role Based Access Control (RBAC)*, a widely used model for access control.

In a centralized system, authorization is based on the closed-world assumption, i.e., all of the parties are known and trusted. In a distributed system, where not all the parties are known *a priori*, the closed-world assumption is not applicable. Trust management systems [43] address the authorization problem in distributed systems by defining a formal language for expressing authorization, and relying on an algorithm to determine when a specific request is allowable. However, as rich languages for expressing security policies emerge, we need to develop techniques to analyze security policies expressed in these languages. In this chapter, we discuss two formalisms for expressing security policies in trust management systems: SPKI/SDSI [103; 106] and RT [160]. We investigate various problems related to these trust management systems using the formalisms and techniques described in Chapter 2.

Analysis of Security Protocols (Chapter 5).   Protocols that enable secure communication over an untrusted network constitute an important part of the current computing infrastructure. Common examples of such protocols are SSL [111], TLS [94], Kerberos [151], and the IPSec [148] and IEEE 802.11i [8] protocol suites. SSL and TLS are used by internet browsers and web servers to allow secure transactions in applications like online banking. The IPSec protocol suite provides confidentiality and integrity at the IP layer and is widely used to secure corporate VPNs. IEEE 802.11i provides data protection and integrity in wireless local area networks, while Kerberos is used for network authentication. The design and security analysis of such network protocols presents a difficult problem. In several instances, serious security vulnerabilities were uncovered in protocols many years after they were first published or deployed [61; 127; 169; 179; 181]. It is therefore critical to have systematic methods for security analysis of such protocols.

The last few years has witnessed much progress in protocol-analysis methods and tools and their application to industrial protocols. This chapter provides a high-level overview of the general methodology of security-protocol analysis. The first step is to define a precise model for representing protocols, adversary capabilities, and the execution of a protocol in conjunction with an adversary. The next step is to define precisely what it means for a protocol to provide a security property, such as *secrecy* and *authentication*, in the face of attack. The final step is to prove that a given protocol actually provides the desired security property, or conversely, to identify attacks on the protocol. Different protocol-analysis frameworks vary in each of these three dimensions. We survey the main lines of work in this area and then focus on a specialized logic, called *Protocol Composition Logic (PCL) [88]*. We illustrate how protocol proofs are carried out in PCL using a signature-based authentication protocol.

In addition, we mention two important problems that have been addressed in recent work. One significant problem has to do with *secure composition of protocols* [88; 117; 130; 170; 249]. Many modern protocols like IKEv2 [147], IEEE 802.11i [8], and Kerberos [151] consist of several different sub-protocols and modes of operation. The challenge is to develop proof methods that allow security proofs of such composite protocols to be built up by combining independent proofs of their parts. Composition is a difficult problem in security since a component may reveal information that does not affect its own security but may degrade the security of some other component in the system.

A second important problem pertains to the *model of protocol execution and attack* used while carrying out the security analysis task. Almost all extant approaches for symbolic protocol analysis use an idealized model in which cryptography is assumed to be perfect. The model developed from positions taken by Needham-Schroeder [201], Dolev-Yao [97], and much subsequent work by others. The idealization makes the protocol-analysis problem more amenable to logical methods. However, the abstraction detracts from the fidelity of the analysis because attacks arising from the interaction between the cryptosystem and the protocol lie outside the scope of the model. The goal then is to develop logical methods for protocol analysis, with associated soundness theorems, that guarantee that a completely symbolic analysis or proof has an interpretation in the standard complexity-theoretic model of modern cryptography [113; 114]. At an informal level, this means that a machine-checkable or generated proof should carry the same meaning as a hand-proof done by a cryptographer. This turns out to be a difficult problem because the security definitions of cryptographic primitives and protocols involve complex probability spaces and quantification over all probabilistic polynomial time attackers. The body of work addressing this problem is referred to as *computationally sound protocol analysis* [12; 15; 33; 90; 91; 165; 183; 192; 193; 212; 219; 220].

CHAPTER 2

Foundations

This chapter discusses foundational concepts and techniques that are used in later chapters. Static analysis is a technique that gathers information about a program without actually executing it on specific inputs. Static analysis is discussed in Section 2.1. Section 2.2 concerns interprocedural dataflow analysis and *weighted pushdown systems (WPDSs)*, a formalism that facilitates interprocedural dataflow analysis. In Chapter 4, we discuss how to use WPDSs to analyze policies in the trust management system SPKI/SDSI. Datalog is a subset of first-order logic, and it is a restricted form of the Prolog logic-programming language. Datalog is discussed in Section 2.3. In Chapter 4, Datalog is used for for analyzing policies in Role Based Access Control and in the trust management system RT.

2.1 STATIC ANALYSIS

2.1.1 WHAT IS STATIC ANALYSIS?

Static analysis provides a way to obtain information about the possible states that a program reaches during execution but without actually running the program on specific inputs. Static-analysis techniques explore the program's behavior for *all* possible inputs and, in principle, account for *all* possible states that the program can reach. In this sense, static analysis is more comprehensive than traditional testing, which tests the program's behavior for a fixed (possibly randomly generated) finite set of runs of the program. However, for any non-trivial program, it is impossible to test explicitly all possible behaviors within a reasonable amount of time; in contrast, static-analysis techniques use *approximations* to account for all of the actions that the program could perform [77]. To make this feasible, two techniques are used:

- The program is *run in the aggregate*. Rather than executing the program on ordinary states, the program is executed on finite-sized descriptors that represent collections of states.

- The program is *run in a non-standard fashion*. Rather than executing the program in a linear sequence, various fragments are executed (in the aggregate) so that, when stitched together, the results are guaranteed to cover all possible execution paths. (Such methods are reviewed in App. A.3.)

For example, one can use descriptors that represent only the sign of a value: neg, zero, pos, and unknown. In a context in which it is known that both a and b are positive (i.e., when the state descriptor is $\langle a \mapsto pos, b \mapsto pos \rangle$), a multiplication expression such "a*b" would be performed as "pos*pos." This approximation discards information about the specific values of a and b: the state descriptor ⟨a ↦ pos, b ↦ pos⟩ represents all states in which a and b hold positive integers.

The choice of which family of descriptors a method uses impacts which behavioral properties of the program can be observed, as well as how efficiently the tool can perform state-space exploration. As reported in [16, p. 719], static analysis was used as early as 1962, and in the intervening forty-seven years researchers have invented new classes of descriptors, as well as new algorithms and heuristics for executing programs on such descriptors, so as to make static analysis useful for larger and larger programs, and to be able to track wider varieties of program properties.

The general goal is to choose a set of descriptors (also called an *abstract domain*) that (i) abstracts away certain details so as to make analysis tractable but (ii) retains enough key information so that the analysis can identify interesting properties that hold. For example, replacing integers with neg, zero, pos, and unknown, restricts each expression to take one of four possible abstract values, whereas the original program could generate an infinite (or very large) number of actual values. However, the abstraction does preserve certain information: because of the properties of multiplication (and abstract multiplication), if a*b evaluates to zero using the abstraction, this guarantees that whenever the expression is evaluated during a run of the program, the expression must always yield 0.

It should be noted that loss of precision is inherent to these techniques, and—especially if an abstraction represents too coarse an approximation—the state-space exploration carried out by the analysis may track behaviors that the original program cannot exhibit. For instance, in the abstract state ⟨a ↦ pos, b ↦ pos⟩, both branches of a test "if (a<=b)" would be explored; however, if the program happened to have the property that a always holds the same value as b at this point, only the true branch would be exercised in any actual execution of the program. Because of this phenomenon, errors reported by an analyzer may be due to error configurations that cannot actually be reached—in other words, they are false positives.

Originally, the motivation for developing static-analysis methods was the desire to improve program performance: information about all states that a program *might* reach (say S) provides information about all states that a program *definitely cannot* reach ($Universe - S$), and this can be used to guide optimizing transformations that, e.g., eliminate code for cases that cannot arise or substitute alternative computations for the cases that can arise (i.e., that are equivalent for states in S, but might not be equivalent for states in $Universe - S$).

The intellectual foundations of static analysis include such classic results as [149] and [77]. Static analysis is covered to some degree in most compiler books, and it has been the subject of several books and monographs [14; 129; 197; 202].

The last nine years have seen much fruitful cross-fertilization between the techniques and ideas developed in the community concerned with static analysis of software and those developed by the hardware-verification community.

Recently, there has also been increased interest in techniques that are *unsound*. Traditionally, research has focused on *sound* techniques that will always report a security vulnerability if one exists.

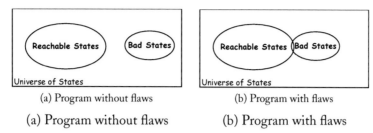

(a) Program without flaws   (b) Program with flaws

Figure 2.1: One use of static analysis is to determine whether situation (a) holds (i.e., the reachable states and the bad states do not overlap) or whether (b) holds (i.e., the reachable states and the bad states do overlap).

Only sound techniques can prove that a class of vulnerabilities is completely absent from a program. However, soundness comes at a cost, and sound techniques are often complex, not very scalable, and overly cautious (they might report many false positives). In contrast, it has recently been shown that unsound techniques can be practical for finding vulnerabilities in large programs, even though they offer no safety guarantees.

2.1.2   HOW IS STATIC ANALYSIS CARRIED OUT?

In this section, we look at some of the ideas that are used in tools for performing static analysis. The goal is to provide insight into what static-analysis tools are capable of achieving, as well as what their limitations are.

The *state* of a program consists of the location of the program counter, plus the values of the program's global variables, stack-allocated variables, and dynamically-allocated memory cells. Given a specification of some set of bad states, the goal of static analysis is to determine which of the situations shown in Fig. 2.1 hold—i.e., whether the reachable states of the program and the bad states overlap or are disjoint. For all but the most trivial programming languages, this property is undecidable; however, this difficulty can often be side-stepped by using approximation techniques. **Abstract Domains**.

Most static-analysis techniques can be viewed as performing *abstract interpretation* [77; 202] of the program. In abstract interpretation, one "executes" the program on a simpler data domain (the *abstract domain*) than the actual data domain (the *concrete domain*) that would be used during a run of the program. The states in the concrete domain involve a huge number of details (e.g., the address of x is 0x7FFF1234, the address of y is 0x7FFF1238, the value of x is 0x7FFF1238, the value of y is 38, etc.). The trick is to devise an abstract domain that eliminates most of these details, yet can retain information about properties of interest. For instance, we might wish to track a fact of the form "x points to y" directly (i.e., as a tuple *points-to*(x, y) in a points-to relation, rather than by representing every detail of the concrete state (i.e., the value of x is 0x7FFF1238, which is the address of y).

Concretization.

In general, elements of the abstract domain ("abstract states") represent *sets* of concrete states so that execution of a statement using an abstract state mimics the execution of the program on all of the concrete states represented. The relationship between abstract states and the corresponding concrete states is captured by a *concretization function*, usually denoted by γ. For instance, one way to track numeric properties is with the abstract domain of intervals [126; 251] in which sets of numeric values (or addresses) are represented using a pair of numbers, which represent a lower and upper bound on the concrete value. Thus, $\gamma([36, 39]) = \{36, 37, 38, 39\}$, and if we have an abstract state $S_0 = \langle a \mapsto [36, 39], b \mapsto [40, 41]\rangle$, then S_0 represents eight concrete states

$$\gamma(S_0) = \{\langle a \mapsto 36, b \mapsto 40\rangle, \langle a \mapsto 36, b \mapsto 41\rangle, \ldots, \langle a \mapsto 39, b \mapsto 40\rangle, \langle a \mapsto 39, b \mapsto 41\rangle, \}.$$

Now suppose that we "execute" the statement a = 2*b; on S_0. Because in interval arithmetic $2 * [40, 41]$ is $[80, 82]$, this results in the abstract state $S_1 = \langle a \mapsto [80, 82], b \mapsto [40, 41]\rangle$, which represents the following six concrete states:

$$\gamma(S_1) = \left\{ \begin{array}{l} \langle a \mapsto 80, b \mapsto 40\rangle, \langle a \mapsto 80, b \mapsto 41\rangle, \\ \langle a \mapsto 81, b \mapsto 40\rangle, \langle a \mapsto 81, b \mapsto 41\rangle, \\ \langle a \mapsto 82, b \mapsto 40\rangle, \langle a \mapsto 82, b \mapsto 41\rangle \end{array} \right\}. \tag{2.1}$$

Note that this is a proper superset of the concrete states that would result from the concrete execution of a = 2*b; on the eight states of $\gamma(S_0)$, which can only produce two of the six states of state-set (2.1), namely,

$$\{\langle a \mapsto 80, b \mapsto 40\rangle, \langle a \mapsto 82, b \mapsto 41\rangle\}. \tag{2.2}$$

In other words, the outcome from abstract execution over-approximates the result from concrete execution. Moreover, an examination of state-set (2.2) reveals that this over-approximation is inherent: state-set (2.2) cannot be represented exactly by an abstract state that consists of an interval for each of a and b. Abstract state S_1 is the most precise abstract state (based on intervals) that over-approximates (2.2).

Joining Abstract States.

Another source of imprecision is that when multiple paths converge, the abstract state is formed by combining the abstract states from each path. This operation, called join (denoted by $\sqcup$), over-approximates set-union. For instance, if one starts in S_0 and executes the fragment

```
if (a >= 38)
    a = 2*b;
/* join point */
```

two abstract states would be joined after the conditional statement: $\langle a \mapsto [80, 82], b \mapsto [40, 41]\rangle$, obtained from the true branch, and $\langle a \mapsto [36, 37], b \mapsto [40, 41]\rangle$, obtained from bypassing the true branch. In the interval domain, the result would be $\langle a \mapsto [36, 82], b \mapsto [40, 41]\rangle$.

Factors that Affect the Choice of Abstract Domain.

```
[1]unsigned int x = 1<<31; // Set x to 2**31
[2]if (trigger) x = 2*x; // Force x to wrap around to 0
[3]if (x == 0) {
[4]   ⟨Do something malicious⟩
[5]}
```

Figure 2.2: The malicious computation in Line 4 would appear to be unreachable to a static-analysis tool that does not account for the properties of two's-complement arithmetic.

The phenomena described above are characteristic of abstract interpretation. Abstract states represent sets of concrete states with some degree of coarseness; each abstract domain has limitations in terms of how precise a result can be obtained.[1] The properties that one is interested in tracking dictate whether an abstract domain is suitable for the problem at hand.

Example 2.1.1 Machine arithmetic is typically 8-bit, 16-bit, 32-bit, or 64-bit two's-complement arithmetic. This means that only a finite set of values can be stored in a variable (or, more precisely, in a machine word or register). Moreover, two's-complement arithmetic is a modular arithmetic: overflow causes values to wrap around.

Failure to model such aspects in the abstract domain can lead to an unsound result (i.e., the result might *not* over-approximate all possible executions). For instance, static analysis based on (infinite-precision) integer arithmetic would be unsound.

To illustrate this, suppose that we wished to carry out an insider attack and hide in a program a malicious computation that could be invoked later under our control. If we knew that a tool based on integer arithmetic (rather than two's-complement arithmetic) would be used to examine the code, we could use the approach shown in Fig. 2.2. In particular, a tool based on integer arithmetic would say that at line [3], x can only have the values 2^{31} and 2^{32}—but never 0. Consequently, it would appear to such a tool that line [4] was unreachable. The result is unsound because the assignment to x in line [2] would give x the value 0, thereby causing line [4] to execute. □

Example 2.1.2 Because on many processors memory accesses do not have to be aligned on word boundaries, the use of the interval domain is problematic for languages in which pointer arithmetic can be performed; an abstract arithmetic based solely on intervals does not provide enough information to check for non-aligned accesses. For example, a 4-byte fetch from memory where the starting address is in the interval [1020, 1028] must be considered to be a fetch from any of the following 4-byte sequences of locations: (1020, . . . , 1023), (1021, . . . , 1024), (1022, . . . , 1025), . . . , (1028, . . . , 1031). Suppose that the program writes the addresses a_1, a_2, and a_3 into the words at (1020, . . . , 1023), (1024, . . . , 1027), and (1028, . . . , 1031), respectively. Because the abstract domain cannot distinguish an unaligned fetch from an aligned fetch, a 4-byte fetch where the starting

---

[1]Other factors that affect the choice of abstract domain are the associated computational costs (i.e., the space to represent abstract states and the time required to perform computations over them).

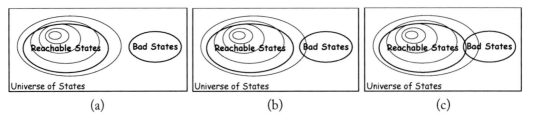

(a)   (b)   (c)

Figure 2.3: After an over-approximation of the reachable states is obtained via successive approximation, three outcomes are possible: (a) Because the apparently-reachable states do not overlap the bad states, over-approximation shows that the reachable states cannot overlap the bad states. (b) The apparently-reachable states overlap the bad states; an error report is a false positive (due to over-approximation). (c) The apparently-reachable states overlap the bad states; an error report could be either a true positive or a false positive. Additional analysis (either manual analysis by a human analyst or a more refined automatic analysis) is needed to distinguish a true positive from a false positive.

address is in the interval [1020, 1028] will appear to allow address forging: e.g., a 4-byte fetch on a little-endian machine from locations $(1021, \ldots, 1024)$ selects the three high-order bytes of a_1, concatenated with the low-order byte of a_2. Moreover, a tool that uses intervals is likely to suffer a catastrophic loss of precision when there are chains of indirection operations: if the first indirection operation fetches the possible values at $(1020, \ldots, 1023)$, $(1021, \ldots, 1024)$, ..., $(1028, \ldots, 1031)$; the second indirection operation will have to follow nine possibilities—including all addresses potentially forged from the sequence a_1, a_2, and a_3. Consequently, the use of intervals in a tool that attempts to identify potential bugs and security vulnerabilities is likely to cause a large number of false alarms to be reported.

In general, to overcome such precision problems it is necessary to use a more precise abstract domain. In this case, we can generalize intervals to *strided intervals* of the form $s[lb, ub]$, which represents the set $\{i \mid lb \leq i \leq ub \wedge \exists j.i = lb + j * s\}$ [213]. (Thus, intervals are strided intervals with strides of 1.)

In the example considered here, if it is known that the starting address of the 4-byte fetch is represented by the strided interval $4[1020, 1028]$, static analysis would discover that the set of possible values is restricted to $\{a_1, a_2, a_3\}$. □

Successive Approximation.

The abstract execution of a program can often be cast as a problem of solving a set of equations by means of successive approximation. If constructed correctly, the execution of the program in the abstract domain over-approximates the semantics of the original program; in particular, any behavior not exhibited when using the abstract domain cannot be exhibited during any program execution (see Fig. 2.3).

2.2   DATAFLOW ANALYSIS, PUSHDOWN SYSTEMS, AND WEIGHTED PUSHDOWN SYSTEMS

Analysis algorithms typically use the program's interprocedural control-flow graph (also known as its *ICFG*). An ICFG consists of a collection of control-flow graphs (CFGs)—one for each procedure—one of which represents the program's main procedure. The CFG for a procedure p has a unique *enter* node and a unique *exit* node. The other nodes represent the program's statements and conditions (or, alternatively, its basic blocks), except that each procedure call in the program is represented in the ICFG by two nodes, a *call* node and a *return-site* node. *Call-edges* connect call nodes to enter nodes; *return-edges* connect exit nodes to return-site nodes. A typical analysis goal is to compute, for each ICFG node n, an overapproximation (i.e., superset) of the set of states that can hold when n is reached.

As already mentioned in §2.1.1, the choice of which family of data descriptors that an algorithm uses impacts which behavioral properties of the program can be observed, including the possible outcome of evaluating a branch-point's condition. This is used to determine an overapproximation of the paths along which control might flow. Thus, a more refined class of data descriptors can sometimes allow certain paths to be excluded from consideration.

On the other hand, certain paths can be excluded merely from consideration of the control-flow properties of the programming language. An important class of paths that can be excluded are those that violate the language's call/return protocol; in particular, an analysis should only consider paths in which the return from a called procedure is matched with the most recent call. Fig. 2.4 shows a fragment of an ICFG, and an example of a path fragment that should be excluded from consideration.

Dataflow-analysis algorithms that exclude such paths have a long history [78; 150; 240]. A natural class of dataflow-analysis problems in which this issue is reduced to a pure graph-reachability problem is also known [214]. The algorithms developed for that class of problems are useful for

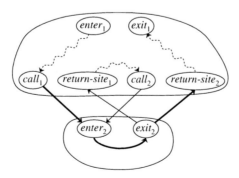

Figure 2.4: An invalid-path fragment: in the path [*call*$_1$, *enter*$_2$, *exit*$_2$, *return-site*$_2$], the return-edge *exit*$_2$ → *return-site*$_2$ does not match with call-edge *call*$_1$ → *enter*$_2$.

analyzing a family of program abstractions called Boolean programs (§2.2.3). (Boolean programs have become well-known due to their use in SLAM [30; 31] to represent program abstractions obtained via predicate abstraction [119].)

More recently, analysis techniques based on pushdown systems (PDSs) [47; 109; 238] have been developed. PDSs are an automata-theoretic formalism for specifying a class of infinite-state transition systems. Infiniteness comes from the fact that each configuration $\langle p, S \rangle$ in the state space consists of a (formal) "control location" p coupled with a stack S of unbounded size. Boolean programs have natural encodings as PDSs (see §2.2.3). Moreover, techniques developed for answering reachability queries on PDSs allow dataflow queries to be posed with respect to a *regular language of configurations*, which allows one to recover dataflow information for specific *calling contexts* (and for regular languages of calling contexts).

Subsequently, these techniques were generalized to *Weighted Pushdown Systems* (WPDSs) [48; 217; 218; 239]. WPDSs extend PDSs by adding a general "black-box" abstraction for expressing transformations of a program's data state (through *weights*). By extending methods from PDSs that answer questions about only certain sets of paths (namely, ones that end in a specified regular language of configurations), WPDSs generalize other frameworks for interprocedural analysis, such as the Sharir-Pnueli functional approach [240], as well as the Knoop-Steffen [150] and Sagiv-Reps-Horwitz summary-based approaches [225]. In particular, conventional dataflow-analysis algorithms merge together the values for all states associated with the same program point, regardless of the states' calling context.

Because WPDSs permit dataflow queries to be posed with respect to a regular language of stack configurations,[2] there are several benefits gained from recasting an existing dataflow-analysis algorithm into the WPDS framework. First, one immediately obtains algorithms to find dataflow information for specific calling contexts and families of calling contexts, which provides information that was not previously obtainable. Second, the algorithms for solving path problems in WPDSs can provide a witness set of paths [218], which is useful for providing an explanation of why the answer to a dataflow query has the value reported.

The remainder of this section is organized as follows: §2.2.1 provides background material on interprocedural dataflow analysis. §2.2.2 defines PDSs, and shows how they can be used to encode (abstractions of) programs. §2.2.3 discusses Boolean programs. §2.2.4 introduces WPDSs, and discusses their use in program analysis.

2.2.1   INTERPROCEDURAL DATAFLOW ANALYSIS

Dataflow analysis is concerned with determining an appropriate dataflow value to associate with each node n in a program, to summarize (safely) some aspect of the possible memory configurations that hold whenever control reaches n. To define an instance of a dataflow problem, one needs

- The CFG of the program.

---

[2]Conventional merged dataflow information can also be obtained by issuing appropriate queries; thus, the new approach provides a strictly richer framework for interprocedural dataflow analysis than prior approaches.

- A join semilattice $(V, \sqcup)$ with least element $\perp$:[3]

  - An element of V represents a set of possible memory configurations. Each point in the program is to be associated with some member of V.
  - The join operator $\sqcup$ is used for combining information obtained along different paths.

- A value $v_0 \in V$ that represents the set of possible memory configurations at the beginning of the program.

- An assignment M of dataflow transfer functions (of type $V \to V$) to the edges of the CFG: $M(e) \in V \to V$.

A dataflow-analysis problem can be formulated as a *path-function problem*.

Definition 2.2.1 *A* **path** *of length $j \geq 1$ from node m to node n is a (non-empty) sequence of j edges, denoted by $[e_1, e_2, \ldots, e_j]$, such that the source of e_1 is m, the target of e_j is n, and for all i, $1 \leq i \leq j - 1$, the target of edge e_i is the source of edge e_{i+1}. For each node m, there is also a path of length 0 from m to m, denoted by $[\ ]_m$ (or by $[\ ]$ when m is understood).* $\square$

The path function pf_q for path $q = [e_1, e_2, \ldots, e_j]$ is the composition, in order, of q's transfer functions: $\mathrm{pf}_q = M(e_j) \circ \ldots \circ M(e_2) \circ M(e_1)$. The path function pf_q for an empty path $q = [\ ]$ is the identity function $\lambda v.v \in V \to V$. In *intra*procedural dataflow analysis, the goal is to determine, for each node n, the "join-over-all-paths" solution:

$$\mathrm{JOP}_n = \bigsqcup_{q \in \mathrm{Paths}(\mathrm{enter},n)} \mathrm{pf}_q(v_0),$$

where $\mathrm{Paths}(\mathrm{enter}, n)$ denotes the set of paths in the CFG from the enter node to n [149]. JOP_n represents a summary of the possible memory configurations that can arise at n: because $v_0 \in V$ represents the set of possible memory configurations at the beginning of the program, $\mathrm{pf}_q(v_0)$ represents the contribution of path q to the memory configurations summarized at n.

The soundness of the JOP_n solution with respect to the programming language's concrete semantics is established by the methodology of *abstract interpretation* [77]:

- A Galois connection (or Galois insertion) is established to define the relationship between sets of concrete states and elements of V.

- Each dataflow transfer function $M(e)$ is shown to overapproximate the transfer function for the concrete semantics of e.

In the discussion below, we assume that such correctness requirements have already been taken care of, and concentrate on algorithms for determining dataflow values once a sound instance of a dataflow-analysis problem has been given.

---

[3]The elements of the semilattice represent program properties. The dataflow-analysis literature [149] and the abstract-interpretation literature [77] differ in the way the semilattice is oriented. Throughout, including Section 2.2.4 on weighted pushdown systems, we will follow the convention used in the abstract-interpretation literature: the operator for combining properties that arise along different paths is $\sqcup$. The reader should be aware that some other presentations of weighted pushdown systems (e.g., [215; 218]) orient the semilattice in the opposite way and use $\sqcap$ to combine properties that arise along different paths.

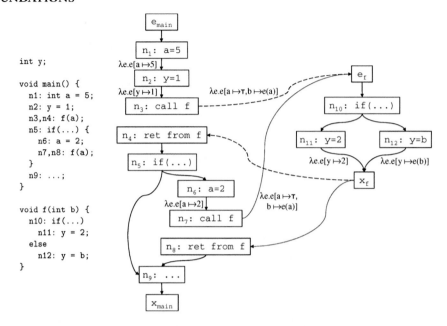

Figure 2.5: A program fragment and its ICFG. For all unlabeled edges, the environment transformer is $\lambda e.e$.

An example ICFG is shown in Fig. 2.5. Let *Var* be the set of all variables in a program, and let $(\mathbb{Z}^\top, \sqsubseteq, \sqcup)$, where $\mathbb{Z}^\top = \mathbb{Z} \cup \{\top\}$, be the standard constant-propagation semilattice: for all $c \in \mathbb{Z}$, $\top \sqsupseteq c$; for all $c_1, c_2 \in \mathbb{Z}$ such that $c_1 \neq c_2$, c_1 and c_2 are incomparable; and $\sqcup$ is the least-upper-bound operation in this partial order. $\top$ stands for "not-a-constant". Let $D = (Env \to Env)$ be the set of all environment transformers where an environment is a mapping for all variables: $Env = (Var \to \mathbb{Z}^\top) \cup \{\bot\}$. We use $\bot$ to denote an infeasible environment. Furthermore, we restrict the set D to contain only $\bot$-strict transformers, i.e., for all $d \in D$, $d(\bot) = \bot$. We can extend the join operation to environments by taking join componentwise.

$$env_1 \sqcup env_2 = \begin{cases} env_1 & \text{if } env_2 = \bot \\ env_2 & \text{if } env_1 = \bot \\ \lambda v.(env_1(v) \sqcup env_2(v)) & \text{otherwise} \end{cases}.$$

The dataflow transformers are shown as edge labels in Fig. 2.5. A transformer of the form $\lambda e.e[a \mapsto 5]$ returns an environment that agrees with the argument, except that a is bound to 5. The environment $\bot$ cannot be updated, and thus $(\lambda e.e[a \mapsto 5])\bot$ equals $\bot$.

The notion of an *(interprocedurally) valid path* captures the idea that not all paths in an ICFG represent potential execution paths. A valid path is one that respects the fact that a procedure always returns to the site of the most recent call. Let each call node in the ICFG be given a unique index from 1 to *CallSites*, where *CallSites* is the total number of call sites in the program. For each call site c_i, label the call-to-enter edge and the exit-to-return-site edge with the symbols "$(_i$" and "$)_i$", respectively. Label all other edges of the ICFG with the symbol e. Each path in the ICFG defines a word, obtained by concatenating—in order—the labels of the edges on the path. A path is a *valid path* iff the path's word is in the language $L(valid)$ generated by the context-free grammar shown below on the left; a path is a *matched path* iff the path's word is in the language $L(matched)$ of balanced-parenthesis strings (interspersed with strings of zero or more e's) generated by the context-free grammar shown below on the right. (In both grammars, i ranges from 1 to *CallSites*.)

$$
\begin{aligned}
valid &\rightarrow matched\ \ valid & matched &\rightarrow matched\ \ matched \\
&\mid\ (_i\ \ valid & &\mid\ (_i\ \ matched\ \ )_i \\
&\mid\ \epsilon & &\mid\ e \\
& & &\mid\ \epsilon
\end{aligned}
$$

The language $L(valid)$ is a language of *partially* balanced parentheses: every right parenthesis "$)_i$" is balanced by a preceding left parenthesis "$(_i$", but the converse need not hold.

Example 2.2.2 In the ICFG shown in Fig. 2.5, the path $[e_{main}, n_1, n_2, n_3, e_f, n_{10}, n_{11}, x_f, n_4, n_5]$ is a matched path, and hence a valid path; the path $[e_{main}, n_1, n_2, n_3, e_f, n_{10}]$ is a valid path, but not a matched path, because the call-to-enter edge $n_3 \rightarrow e_f$ has no matching exit-to-return-site edge; the path $[e_{main}, n_1, n_2, n_3, e_f, n_{10}, n_{11}, x_f, n_8]$ is neither a matched path nor a valid path because the exit-to-return-site edge $x_f \rightarrow n_8$ does not correspond to the preceding call-to-enter edge $n_3 \rightarrow e_f$.
□

In interprocedural dataflow analysis, the goal shifts from finding the join-over-*all*-paths solution to the more precise "join-over-*all–valid*-paths", or "context-sensitive" solution. A context-sensitive interprocedural dataflow analysis is one in which the analysis of a called procedure is "sensitive" to the context in which it is called. A context-sensitive analysis captures the fact that the results propagated back to each return site r should depend only on the memory configurations that arise at the call site that corresponds to r. More precisely, the goal of a context-sensitive analysis is to find the join-over-all-valid-paths value for nodes of the ICFG [150; 225; 240]:

$$
\text{JOVP}_n = \bigsqcup_{q \in \text{VPaths}(e_{main}, n)} \text{pf}_q(v_0),
$$

where VPaths(e_{main}, n) denotes the set of valid paths from the main procedure's enter node to n.
Although some valid paths may also be infeasible execution paths, none of the non-valid paths are feasible execution paths. By restricting attention to just the valid paths from e_{main}, we thereby exclude some of the infeasible execution paths. In general, therefore, JOVP_n characterizes the memory configurations at n more precisely than JOP_n.

2.2.2 PUSHDOWN SYSTEMS
In this section, we define pushdown systems and show how they can be used to encode ICFGs.

Rule	Control flow modeled
$\langle p, u \rangle \hookrightarrow \langle p, v \rangle$	Intraprocedural edge $u \to v$
$\langle p, c \rangle \hookrightarrow \langle p, e_f\, r \rangle$	Call to f from c that returns to r
$\langle p, x_f \rangle \hookrightarrow \langle p, \varepsilon \rangle$	Return from f at exit node x_f

Figure 2.6: The encoding of an ICFG's edges as PDS rules.

Definition 2.2.3 *A **pushdown system** is a triple $\mathcal{P} = (P, \Gamma, \Delta)$, where P is a finite set of states (also known as "control locations"), Γ is a finite set of stack symbols, and $\Delta \subseteq P \times \Gamma \times P \times \Gamma^*$ is a finite set of rules. A **configuration** of $\mathcal{P}$ is a pair $\langle p, u \rangle$ where $p \in P$ and $u \in \Gamma^*$. A rule $r \in \Delta$ is written as $\langle p, \gamma \rangle \hookrightarrow \langle p', u \rangle$, where $p, p' \in P$, $\gamma \in \Gamma$ and $u \in \Gamma^*$. These rules define a transition relation $\Rightarrow$ on configurations of $\mathcal{P}$ as follows: If $r = \langle p, \gamma \rangle \hookrightarrow \langle p', u' \rangle$, then $\langle p, \gamma u \rangle \Rightarrow \langle p', u'u \rangle$ for all $u \in \Gamma^*$. The reflexive transitive closure of $\Rightarrow$ is denoted by $\Rightarrow^*$. For a set of configurations C, we define $pre^*(C) = \{c' \mid \exists c \in C : c' \Rightarrow^* c\}$ and $post^*(C) = \{c' \mid \exists c \in C : c \Rightarrow^* c'\}$, which are just backward and forward reachability under the transition relation $\Rightarrow$.* □

Without loss of generality, we restrict the pushdown rules to have at most two stack symbols on the right-hand side [238]. A rule $r = \langle p, \gamma \rangle \hookrightarrow \langle p', u \rangle$, $u \in \Gamma^*$, is called a *pop* rule if $|u| = 0$, and a *push* rule if $|u| = 2$.

The PDS configurations model (node, stack) pairs of the program's state. Given a program P, we can use a PDS to model a limited portion of a P's behavior in the following sense: the configurations of the PDS represent a superset of P's (node, stack) pairs.

The standard approach for modeling a program's control flow with a pushdown system is as follows: P contains a single state p, Γ corresponds to the nodes of the program's ICFG, and Δ corresponds to edges of the program's ICFG (see Fig. 2.6). For instance, the rules that encode the ICFG shown in Fig. 2.5 are

$$
\begin{array}{lll}
\langle p, e_{main} \rangle \hookrightarrow \langle p, n_1 \rangle & \langle p, n_5 \rangle \hookrightarrow \langle p, n_9 \rangle & \langle p, e_f \rangle \hookrightarrow \langle p, n_{10} \rangle \\
\langle p, n_1 \rangle \hookrightarrow \langle p, n_2 \rangle & \langle p, n_6 \rangle \hookrightarrow \langle p, n_7 \rangle & \langle p, n_{10} \rangle \hookrightarrow \langle p, n_{11} \rangle \\
\langle p, n_2 \rangle \hookrightarrow \langle p, n_3 \rangle & \langle p, n_7 \rangle \hookrightarrow \langle p, e_f\, n_8 \rangle & \langle p, n_{11} \rangle \hookrightarrow \langle p, x_f \rangle \\
\langle p, n_3 \rangle \hookrightarrow \langle p, e_f\, n_4 \rangle & \langle p, n_8 \rangle \hookrightarrow \langle p, n_9 \rangle & \langle p, n_{10} \rangle \hookrightarrow \langle p, n_{12} \rangle \\
\langle p, n_4 \rangle \hookrightarrow \langle p, n_5 \rangle & \langle p, n_9 \rangle \hookrightarrow \langle p, x_{main} \rangle & \langle p, n_{12} \rangle \hookrightarrow \langle p, x_f \rangle \\
\langle p, n_5 \rangle \hookrightarrow \langle p, n_6 \rangle & \langle p, x_{main} \rangle \hookrightarrow \langle p, \varepsilon \rangle & \langle p, x_f \rangle \hookrightarrow \langle p, \varepsilon \rangle
\end{array}
$$

PDSs that have only a single control location, as discussed above, are also called "context-free processes" [55]. In §2.2.3, we will discuss how, in addition to control flow, PDSs can also be used to encode program models that involve finite abstractions of the program's data. PDSs that have multiple control locations are used in such encodings.

The problem of interest is to find the set of all reachable configurations, starting from a given set of configurations. This can then be used, for example, for assertion checking (i.e., determining if a given assertion can ever fail) or to find the set of all data values that may arise at a program point (for dataflow analysis).

Because the number of configurations of a pushdown system is unbounded, it is useful to use finite automata to describe regular sets of configurations.

Definition 2.2.4 *If $\mathcal{P} = (P, \Gamma, \Delta)$ is a PDS then a $\mathcal{P}$-**automaton** is a finite automaton $(Q, \Gamma, \rightarrow , P, F)$, where $Q \supseteq P$ is a finite set of states, $\rightarrow \subseteq Q \times \Gamma \times Q$ is the transition relation, P is the set of initial states, and F is the set of final states. We say that a configuration $\langle p, u \rangle$ is accepted by a $\mathcal{P}$-automaton if the automaton can accept u when it is started in the state p (written as $p \xrightarrow{u}{}^{*} q$, where $q \in F$). A set of configurations is called **regular** if some $\mathcal{P}$-automaton accepts it. Without loss of generality, $\mathcal{P}$-automata are restricted to not have any transitions leading to an initial state.* □

An important result is that for a regular set of configurations C, both $post^{*}(C)$ and $pre^{*}(C)$ (the forward and the backward reachable sets of configurations, respectively) are also regular sets of configurations [47; 53]. The algorithms for computing $post^{*}$ and pre^{*}, called *poststar* and *prestar*, respectively, take a $\mathcal{P}$-automaton $\mathcal{A}$ as input, and if C is the set of configurations accepted by $\mathcal{A}$, they produce $\mathcal{P}$-automata $\mathcal{A}_{post^{*}}$ and $\mathcal{A}_{pre^{*}}$ that accept the sets of configurations $post^{*}(C)$ and $pre^{*}(C)$, respectively [47; 107; 109]. Both *poststar* and *prestar* can be implemented as *saturation procedures*; i.e., transitions are added to $\mathcal{A}$ according to some saturation rule until no more can be added.

Algorithm *prestar*. $\mathcal{A}_{pre^{*}}$ can be constructed from $\mathcal{A}$ using the following saturation rule: *If $\langle p, \gamma \rangle \hookrightarrow \langle p', w \rangle$ and $p' \xrightarrow{w} q$ in the current automaton, add a transition (p, γ, q).*

Algorithm *poststar*. $\mathcal{A}_{post^{*}}$ can be constructed from $\mathcal{A}$ by performing Phase I and then saturating via the rules given in Phase II:

- *Phase I.* For each pair (p', γ') such that $\mathcal{P}$ contains at least one rule of the form $\langle p, \gamma \rangle \hookrightarrow \langle p', \gamma'\gamma'' \rangle$, add a new state $p'_{\gamma'}$.

- *Phase II (saturation phase).* (The symbol $\xrightarrow{\gamma}$ denotes the relation $(\xrightarrow{\epsilon})^{\star} \xrightarrow{\gamma} (\xrightarrow{\epsilon})^{\star}$.)

    - If $\langle p, \gamma \rangle \hookrightarrow \langle p', \epsilon \rangle \in \Delta$ and $p \xrightarrow{\gamma} q$ in the current automaton, add a transition (p', ϵ, q).

    - If $\langle p, \gamma \rangle \hookrightarrow \langle p', \gamma' \rangle \in \Delta$ and $p \xrightarrow{\gamma} q$ in the current automaton, add a transition (p', γ', q).

    - If $\langle p, \gamma \rangle \hookrightarrow \langle p', \gamma'\gamma'' \rangle \in \Delta$ and $p \xrightarrow{\gamma} q$ in the current automaton, add the transitions $(p', \gamma', p'_{\gamma'})$ and $(p'_{\gamma'}, \gamma'', q)$.

Example 2.2.5 Given the PDS that encodes the ICFG from Fig. 2.5 and the query automaton $\mathcal{A}$ shown in Fig. 2.7(a), which accepts the language $\{\langle p, e_{main} \rangle\}$, *poststar* produces the automaton $\mathcal{A}_{post^{*}}$ shown in Fig. 2.7(b), □

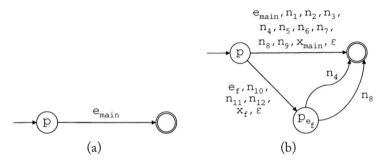

(a)   (b)

Figure 2.7: (a) Automaton for the input language of configurations $\{\langle p, e_{main}\rangle\}$; (b) automaton for $post^*(\{\langle p, e_{main}\rangle\})$ (computed for the PDS that encodes the ICFG from the constant propagation example.

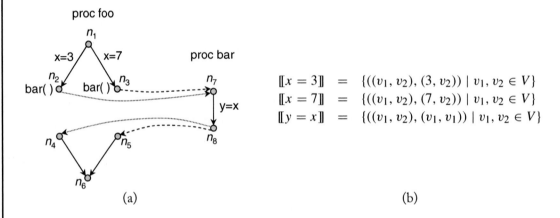

(a)   (b)

Figure 2.8: (a) A Boolean program with two procedures and two global variables x and y over a finite domain $V = \{0, 1, \ldots, 7\}$. (b) The (non-identity) transformers used in the Boolean program.

2.2.3   BOOLEAN PROGRAMS

A Boolean program can be thought of as a C program that uses only the Boolean datatype and fixed-length vectors of Booleans. It does not have any pointers or heap-allocated storage. A Boolean program consists of a finite set of procedures. It has a finite set of global variables, and a finite set of local variables for each procedure. Each variable can only hold a value from a finite domain.[4] To simplify the discussion, we assume that procedures do not have parameters (they can be passed through global variables). The variables in scope inside a procedure are the global variables and its set of local variables. Fig. 2.8(a) shows a Boolean program with two procedures and two global variables x and y over a finite domain $V = \{0, 1, \ldots, 7\}$.

---

[4]An assignment to a variable v that holds a value from a finite domain can be thought of as a collection of assignments to a *vector* of Boolean-valued variables, namely, the collection of Boolean-valued variables that holds the encoding of v's value.

Notation. A binary relation on a set S is a subset of $S \times S$. If R_1 and R_2 are binary relations on S, then their relational composition, denoted by "$R_1; R_2$", is defined by $\{(s_1, s_3) \mid \exists s_2 \in S, (s_1, s_2) \in R_1, (s_2, s_3) \in R_2\}$. If R is a binary relation, R^i is the relational composition of R with itself i times, and R^0 is the identity relation on S. $R^* = \cup_{i=0}^{\infty} R^i$ is the reflexive-transitive closure of R.

Let G be the set of valuations of the global variables, and let Val_i be the set of valuations of the local variables of procedure i. Let L be the set of local states of the program; each local state consists of the value of the program counter, a valuation of local variables from some Val_i, and the program stack (which, for each unfinished call to a procedure P, contains a return address and a valuation of the local variables of P).

The effect of executing an assignment or assume statement st, denoted by $[\![\mathtt{st}]\!]$, is a binary relation on $G \times \mathrm{Val}_i$ that describes how values of variables in scope can change. Fig. 2.8(b) shows the (non-identity) transformers used in Fig. 2.8(a).

To encode a Boolean program using a PDS, the state alphabet P is expanded to encode the values of global variables, and the stack alphabet is expanded to encode the values of local variables [238].

Let N_i be the set of control locations of the ith procedure. We set P to be G, and Γ to be the union of $N_i \times \mathrm{Val}_i$ over all procedures. (Note that the set of local states L equals Γ^*.) The PDS rules for the ith procedure are constructed as follows: (i) an intraprocedural ICFG edge $u \to v$ with action st is encoded via a set of rules $\langle g, (u, l) \rangle \hookrightarrow \langle g', (v, l') \rangle$, for each $((g, l), (g', l')) \in [\![\mathtt{st}]\!]$; (ii) a call edge $c \to r$ that calls procedure f, with enter node e_f, is encoded via a set of rules $\langle g, (c, l) \rangle \hookrightarrow \langle g, (e_f, l_0) (r, l) \rangle$, for each $(g, l) \in G \times \mathrm{Val}_i$ and $l_0 \in \mathrm{Val}_f$; (iii) a procedure return at node u is encoded via a set of rules $\langle g, (u, l) \rangle \hookrightarrow \langle g, \varepsilon \rangle$, for each $(g, l) \in G \times \mathrm{Val}_i$.

Under such an encoding of a Boolean program as a PDS, a configuration $\langle p, \gamma_1 \gamma_2 \cdots \gamma_n \rangle$ is an element of $G \times L$ that describes the instantaneous state of a program. The state p encodes the values of global variables; γ_1 encodes the current program location and the values of local variables in scope; and the rest of the stack encodes the list of unfinished calls with the values of local variables at the time the call was made. The PDS transition relation ($\Rightarrow$), which is essentially a transition relation on $G \times L$, represents the semantics of the Boolean program.

2.2.4   WEIGHTED PUSHDOWN SYSTEMS

A weighted pushdown system is obtained by augmenting a PDS with a weight domain that is a *bounded idempotent semiring* [48; 218]. Such semirings are powerful enough to encode finite-state data abstractions, such as the ones required for bitvector dataflow analysis, Boolean programs, and the IFDS framework of Reps et al. [214], as well as infinite-state data abstractions, such as linear-constant propagation [225] and affine-relation analysis [198; 199]. We present some of this here; additional material about using WPDSs for interprocedural analysis can be found in [218].

Weights encode the effect that each statement (or PDS rule) has on the data state of the program. They can be thought of as abstract transformers that specify how the abstract state changes when a statement is executed.

Definition 2.2.6 *A* **bounded idempotent semiring** *(or* **weight domain***) is a tuple* $(D, \oplus, \otimes, \bar{0}, \bar{1})$, *where D is a set whose elements are called* **weights**, $\bar{0}, \bar{1} \in D$, *and* $\oplus$ *(the combine operation) and* $\otimes$ *(the extend operation) are binary operators on D such that*

1. $(D, \oplus)$ *is a commutative monoid with* $\bar{0}$ *as its neutral element, and where* $\oplus$ *is idempotent.* $(D, \otimes)$ *is a monoid with the neutral element* $\bar{1}$.

2. $\otimes$ *distributes over* $\oplus$, *i.e., for all* $a, b, c \in D$ *we have*
$$a \otimes (b \oplus c) = (a \otimes b) \oplus (a \otimes c) \text{ and } (a \oplus b) \otimes c = (a \otimes c) \oplus (b \otimes c) .$$

3. $\bar{0}$ *is an annihilator with respect to* $\otimes$, *i.e., for all* $a \in D$, $a \otimes \bar{0} = \bar{0} = \bar{0} \otimes a$.

4. *In the partial order* $\sqsubseteq$ *defined by* $\forall a, b \in D$, $a \sqsubseteq b$ *iff* $a \oplus b = b$, *there are no infinite strictly ascending chains.*

□

Definition 2.2.7 *A* **weighted pushdown system** *is a triple* $W = (\mathcal{P}, \mathcal{S}, f)$, *where* $\mathcal{P} = (P, \Gamma, \Delta)$ *is a PDS,* $\mathcal{S} = (D, \oplus, \otimes, \bar{0}, \bar{1})$ *is a bounded idempotent semiring, and* $f : \Delta \rightarrow D$ *is a map that assigns a weight to each rule of* $\mathcal{P}$. □

WPDSs compute over the weights via the extend operation ($\otimes$). Let $\sigma \in \Delta^*$ be a sequence of rules. Using f, we can associate a value to σ; i.e., if $\sigma = [r_1, \ldots, r_k]$, we define $v(\sigma) \overset{\text{def}}{=} f(r_1) \otimes \ldots \otimes f(r_k)$. In program-analysis problems, weights typically represent abstract transformers that specify how the abstract state changes when a statement is executed. Thus, the extend operation is typically the reversal of function composition: $w_1 \otimes w_2 = w_2 \circ w_1$. (Computing over transformers by composing them—instead of computing on the underlying abstract states by applying transformers to abstract states—is customary in interprocedural analysis, where procedure summaries need to be calculated as compositions of abstract-state transformers [78; 150; 214].)

Reachability problems on PDSs are generalized to WPDSs as follows:

Definition 2.2.8 *Let* $W = (\mathcal{P}, \mathcal{S}, f)$ *be a weighted pushdown system, where* $\mathcal{P} = (P, \Gamma, \Delta)$. *For any two configurations c and c' of* $\mathcal{P}$, *let* $path(c, c')$ *denote the set of all rule sequences that transform c into c'. Let* $S, T \subseteq P \times \Gamma^*$ *be regular sets of configurations. If* $\sigma \in path(c, c')$, *then we say* $c \Rightarrow^{\sigma} c'$. *The* **join-over-all-valid-paths** *value* $JOVP(S, T)$ *is defined as* $\bigoplus \{v(\sigma) \mid s \Rightarrow^{\sigma} t, s \in S, t \in T\}$. □

A PDS, as defined in §2.2.2, is simply a WPDS with the *Boolean weight domain* $(\{F, T\}, \vee, \wedge, F, T)$ and weight assignment $f(r) = T$ for all rules $r \in \Delta$. In this case, $JOVP(S, U) = T$ iff there exists a path from a configuration in S to a configuration in U, i.e., $post^*(S) \cap U$ and $S \cap pre^*(U)$ are non-empty sets.

One way of modeling a program as a WPDS is as follows: the PDS models the control flow of the program, as in Fig. 2.6. The weight domain models abstract transformers for an abstraction of the program's data. §2.2.4.1 and §2.2.4.2 describe several data abstractions that can be encoded

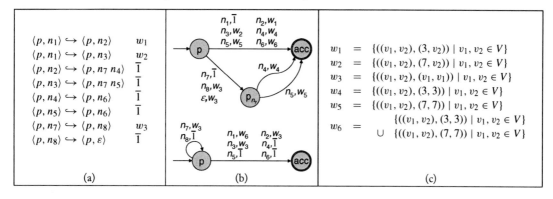

Figure 2.9: (a) A WPDS that encodes the Boolean program from the previous figure. (b) The result of *poststar*($\langle p, n_1\rangle$) and *prestar*($\langle p, n_6\rangle$). The final state in each of the automata is *acc*. (c) Definitions of the weights used in the figure.

using weight domains. To simplify the presentation, we only show the treatment for global variables, and do not consider local variables. Finite-state abstractions of local variables can always be encoded in the stack alphabet, as for PDSs [152; 238]. For infinite-state abstractions, local variables pose an extra complication for WPDSs [152]; their treatment is discussed in §2.2.4.4.

2.2.4.1 Finite-State Data Abstractions
An important weight domain for WPDSs is the set of all binary relations on a finite set.

Definition 2.2.9 *If G is a finite set, then the* **relational weight domain** *on G is defined as* $(2^{G \times G}, \cup, ;, \emptyset, id)$: *weights are binary relations on G, combine is union, extend is relational composition (";"), $\overline{0}$ is the empty relation, and $\overline{1}$ is the identity relation on G.* □

By instantiating G to be the set of global states of a Boolean program P,[5] we obtain a weight domain for encoding P. This approach yields a more straightforward encoding of P: the weight associated with the rule that encodes an assignment or assume statement st of P is exactly [[st]]— i.e., its effect on the global state of P—which, as described in §2.2.3, is a binary relation on G. For example, the WPDS shown in Fig. 2.9 encodes the Boolean program from Fig. 2.8(a). The Boolean program has two variables that range over the set $V = \{0, 1, \ldots, 7\}$, so $G = V \times V$, where the two components represent the values of x and y, respectively.

The set of all data values that reach a node n can be calculated as follows: let S be the singleton configuration consisting of the program's enter node, $S = \{\langle p, e_{main}\rangle\}$, and let T be the set $\{\langle p, n\, u\rangle \mid u \in \Gamma^*\}$. Let $w = \text{JOVP}(S, T)$. If $w = \overline{0}$, then the node cannot be reached. Otherwise, w captures the net transformation on the global state from when the program started. The range of w, i.e., the set $\{g \in G \mid \exists g' \in G : (g', g) \in w\}$, is the set of valuations that reach node n. For

---
[5] A *global state* of P is a valuation of P's global variables.

example, in Fig. 2.9, the JOVP weight to node n_6 is the weight w_6 shown in Fig. 2.9(c). Its range shows that either $x = 3$ and $y = 3$, or $x = 7$ and $y = 7$.

Because T can be any regular set, one can also answer stack-qualified queries [218]. For example, the set of values that arise at node n when its procedure is called from call site m can be found by setting $T = \{\langle p, n\, m_r\, u \rangle \mid u \in \Gamma^*\}$, where m_r is the return site for call site m.

A WPDS with a weight domain that has a finite set of weights, such as the one described above, can also be encoded as a PDS. However, it is often useful to use weights because they can be symbolically encoded. Tools such as MOPED and SLAM use BDDs [52] to encode sets of data values, which allows them to scale to a large number of variables. (Using PDSs for Boolean program verification, without any symbolic encoding, is generally not a feasible approach.)

2.2.4.2 Infinite-State Data Abstractions

An infinite-state data abstraction is one in which the number of abstract states (or weights) is infinite. We begin with two simple examples of infinite weight domains, and then discuss the weight domain used for affine-relation analysis.

Finding Shortest Valid Paths.

Definition 2.2.10 *The* **minpath semiring** *is the weight domain* $\mathcal{M} = (\mathbb{N} \cup \{\infty\}, min, +, \infty, 0)$: *weights are non-negative integers including "infinity", combine is minimum, and extend is addition.* □

If all rules of a WPDS are given the weight $1 \in \mathbb{N}$ from this semiring (different from the semiring weight $\bar{1}$, which is the integer $0 \in \mathbb{N}$), then the JOVP weight between two configurations is the length of the shortest path (shortest rule sequence) between them.

Finding Shortest Traces.

The minpath semiring can be combined with a relational weight domain, for example, to find the shortest (valid) path in a Boolean program (for finding the shortest trace that exhibits some property).

Definition 2.2.11 *A* **weighted relation** *on a set S, weighted with semiring $(D, \oplus, \otimes, \bar{0}, \bar{1})$, is a function from $(S \times S)$ to D. The composition of two weighted relations R_1 and R_2 is defined as $(R_1; R_2)(s_1, s_3) = \oplus\{w_1 \otimes w_2 \mid \exists s_2 \in S : w_1 = R_1(s_1, s_2), w_2 = R_2(s_2, s_3)\}$. The union of the two weighted relations is defined as $(R_1 \cup R_2)(s_1, s_2) = R_1(s_1, s_2) \oplus R_2(s_1, s_2)$. The identity relation is the function that maps each pair (s, s) to $\bar{1}$ and others to $\bar{0}$. The reflexive transitive closure is defined in terms of these operations, as before. If $\rightarrow$ is a weighted relation and $(s_1, s_2, w) \in \rightarrow$, then we write $s_1 \xrightarrow{w} s_2$.* □

Definition 2.2.12 *If S is a weight domain with set of weights D and G is a finite set, then the* **relational weight domain** *on (G, S) is defined as $(2^{G \times G \rightarrow D}, \cup, ;, \emptyset, id)$: weights are weighted relations on G and the operations are the corresponding ones for weighted relations.* □

If G is the set of global states of a Boolean program, then the relational weight domain on $(G, \mathcal{M})$ can be used for finding the shortest trace: for each rule, if $R \subseteq G \times G$ is the effect of

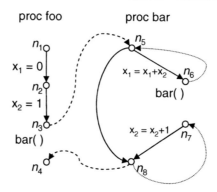

Figure 2.10: An affine program that starts execution at node n_1. There are two global variables x_1 and x_2.

executing the rule on the global state of the Boolean program, then associate the rule with the following weight:

$$\{g_1 \xrightarrow{1} g_2 \mid (g_1, g_2) \in R\} \cup \{g_1 \xrightarrow{\infty} g_2 \mid (g_1, g_2) \notin R\}.$$

Then, if $w = \text{JOVP}(C_1, C_2)$, the length of the shortest path that starts with global state g from a configuration in C_1 and ends at global state g' in a configuration in C_2, is $w(g, g')$ (which would be ∞ if no path exists). (Moreover, if a finite-length path does exist, a witness trace [218] can be obtained to identify the elements of the path.)

Affine-Relation Analysis.

An affine relation is a linear-equality constraint between integer-valued variables. Affine-relation analysis (ARA) tries to find all affine relationships that hold in the program. An example is shown in Fig. 2.10. For this program, ARA would infer that $x_2 = x_1 + 1$ always holds at program node n_4.

ARA for single-procedure programs was first addressed by Karr [146]. ARA generalizes other analyses, including copy-constant propagation, linear-constant propagation [225], and induction-variable analysis [146].

Affine Programs.   Interprocedural ARA can be performed precisely on *affine programs*, and has been the focus of several papers [122; 198; 199]. Affine programs are similar to Boolean programs, but with integer-valued variables. Again, we restrict our attention to global variables and defer treatment of local variables to §2.2.4.4. If $\{x_1, x_2, \cdots, x_n\}$ is the set of global variables of the program, then all assignments have the form $x_j := a_0 + \sum_{i=1}^{n} a_i x_i$, where $a_0, \cdots, a_n$ are integer constants. An assignment can also be non-deterministic, denoted by $x_j := ?$, which may assign any integer to x_j. (This is typically used for abstracting assignments that cannot be modeled as an affine transformation of the variables.) All branch conditions in affine programs are non-deterministic.

ARA Weight Domain. We briefly describe the weight domain based on the linear-algebra formulation of ARA from [198]. An affine relation $a_0 + \sum_{i=1}^{n} a_i x_i = 0$ is represented using a column vector of size $n + 1$: $\vec{a} = (a_0, a_1, \cdots, a_n)^t$. A valuation of program variables $\bar{x}$ is a map from the set of global variables to the integers. The value of x_i under this valuation is written as $\bar{x}(i)$.

A valuation $\bar{x}$ satisfies an affine relation $\vec{a} = (a_0, a_1, \cdots, a_n)^t$ if $a_0 + \sum_{i=1}^{n} a_i \bar{x}(i) = 0$. An affine relation $\vec{a}$ represents the set of all valuations that satisfy it, written as $Pts(\vec{a})$. An affine relation $\vec{a}$ holds at a program node if the set of valuations reaching that node (in the concrete collecting semantics) is a subset of $Pts(\vec{a})$.

An important observation about affine programs is that if affine relations $\vec{a}_1$ and $\vec{a}_2$ hold at a program node, then so does any linear combination of $\vec{a}_1$ and $\vec{a}_2$. For example, one can verify that $Pts(\vec{a}_1 + \vec{a}_2) \supseteq Pts(\vec{a}_1) \cap Pts(\vec{a}_2)$, i.e., the affine relation $\vec{a}_1 + \vec{a}_2$ (componentwise addition) holds at a program node if both $\vec{a}_1$ and $\vec{a}_2$ hold at that node. The set of affine relations that hold at a program node forms a (finite-dimensional) vector space [198]. This implies that a (possibly infinite) set of affine relations can be represented by any of its bases; each such basis is always a finite set.

For reasoning about affine programs, Müller-Olm and Seidl defined an abstraction that is able to find all affine relationships in an affine program: each statement is abstracted by a set of matrices of size $(n + 1) \times (n + 1)$. This set is the weakest-precondition transformer on affine relations for that statement: if a statement is abstracted as the set $\{m_1, m_2, \cdots, m_r\}$, then the affine relation $\vec{a}$ holds after the execution of the statement if and only if the affine relations $(m_1 \vec{a}), (m_2 \vec{a}), \cdots, (m_r \vec{a})$ all hold before the execution of the statement.

Under such an abstraction of program statements, one can define the extend operation, which is transformer composition, as elementwise matrix multiplication, and the combine operation as set union. This is correct semantically, but it does not give an effective algorithm because the matrix sets can grow unboundedly. However, the observation that affine relations form a vector space carries over to a set of matrices as well. One can show that the transformer $\{m_1, m_2, \cdots, m_r\}$ is semantically equivalent to the transformer $\{m_1, m_2, \cdots, m_r, m\}$, where m is any linear combination of the m_i matrices. Thus, a set of matrices can be abstracted as the (infinite) set of matrices spanned by them. Once we have a vector space, we can represent it using any of its bases to get a finite and bounded representation: a vector space over matrices of size $(n + 1) \times (n + 1)$ cannot have more that $(n + 1)^2$ matrices in any basis.

If M is a set of matrices, let $Span(M)$ be the vector space spanned by them. Let β be the basis operation that takes a set of matrices and returns a basis of their span. We can now define the weight domain. A weight w is a vector space of matrices, which can be represented using its basis. Extend of vector spaces w_1 and w_2 is the vector space $\{(m_1 m_2) \mid m_i \in w_i\}$. Combine of w_1 and w_2 is the vector space $\{(m_1 + m_2) \mid m_i \in w_i\}$, which is the smallest vector space containing both w_1 and w_2. $\bar{0}$ is the empty set, and $\bar{1}$ is the span of the singleton set consisting of the identity matrix. The extend and combine operations, as defined above, are operations on infinite sets. They can be implemented by the corresponding operations on any basis of the weights. The following properties show that it is semantically correct to operate on the elements in the basis instead of all the elements in the vector space spanned by them:

$$\beta(w_1 \oplus w_2) = \beta(\beta(w_1) \oplus \beta(w_2))$$
$$\beta(w_1 \otimes w_2) = \beta(\beta(w_1) \otimes \beta(w_2))$$

These properties are satisfied because of the linearity of extend (matrix multiplication distributes over addition) and combine operations.

Under such a weight domain, JOVP(S, T) is a weight that is the net weakest-precondition transformer between S and T. Suppose that this weight has the basis $\{m_1, \cdots, m_r\}$. The affine relation that indicates that any variable valuation might hold at S is $\vec{0} = (0, 0, \cdots, 0)$. Thus, $\vec{0}$ holds at S, and the affine relation $\vec{a}$ holds at T iff $m_1\vec{a} = m_2\vec{a} = \cdots = m_r\vec{a} = \vec{0}$. The set of all affine relations that hold at T can be found as the intersection of the null spaces of the matrices $m_1, m_2, \cdots, m_r$.

Extensions to ARA.   ARA can also be performed for modular arithmetic [199] to precisely model machine arithmetic (which is modulo 2 to the power of the word size). The weight domain is similar to the one described above.

2.2.4.3 Solving for the JOVP Value

There are two algorithms for solving for JOVP values, called *prestar* and *poststar* (by analogy with the algorithms for PDSs). They take as input an automaton that accepts the set of initial configurations. As output, they produce a *weighted automaton*:

Definition 2.2.13 *Given a weighted pushdown system $\mathcal{W} = (\mathcal{P}, \mathcal{S}, f)$, a $\mathcal{W}$-**automaton** $\mathcal{A}$ is a $\mathcal{P}$-automaton, where each transition in the automaton is labeled with a weight. The weight of a path in the automaton is obtained by taking an extend of the weights on the transitions in the path in either a forward or backward direction. The automaton is said to accept a configuration $c = \langle p, u \rangle$ with weight $w = \mathcal{A}(c)$ if w is the combine of weights of all accepting paths for u starting from state p in $\mathcal{A}$. We call the automaton a **backward $\mathcal{W}$-automaton** if the weight of a path is read backwards, and a **forward $\mathcal{W}$-automaton** otherwise.* □

Let $\mathcal{A}$ be an unweighted automaton and $\mathcal{L}(\mathcal{A})$ be the set of configurations accepted by it. Then, *prestar*($\mathcal{A}$) produces a forward weighted automaton $\mathcal{A}_{pre^*}$ as output, such that $\mathcal{A}_{pre^*}(c) =$ JOVP($\{c\}, \mathcal{L}(\mathcal{A})$), whereas *poststar*($\mathcal{A}$) produces a backward weighted automaton $\mathcal{A}_{post^*}$ as output, such that $\mathcal{A}_{post^*}(c) = $JOVP($\mathcal{L}(\mathcal{A}), \{c\}$) [218]. Examples are shown in Fig. 2.9(b). One thing to note here is how the *poststar* automaton works. The procedure bar is analyzed independently of its calling context (i.e., without knowing the exact value of x), which generates the transitions between p and p_{n_7}. The calling context of bar, which determines the input values to bar, is represented by the transitions that leave state p_{n_7}. This is how, for instance, the automaton records that x = 3 and y = 3 at node n_8 when bar is called from node n_2 (in which case the return-site node on the top of the stack is n_4).

Using standard automata-theoretic techniques, one can also compute $\mathcal{A}_w(C)$ for (forward or backward) weighted automaton $\mathcal{A}_w$ and a regular set of configurations C, where $\mathcal{A}_w(C) = \bigoplus \{\mathcal{A}_w(c) \mid c \in C\}$. This allows one to solve for the join-over-all-paths value $\text{JOVP}(S, T)$ for configuration sets S and T by computing either $poststar(S)(T)$ or $prestar(T)(S)$.

We briefly describe how the *prestar* algorithm works for WPDSs. The interested reader is referred to [218] for more details (e.g., the *poststar* algorithm), as well as an efficient implementation of the algorithm. The algorithm takes an unweighted automaton $\mathcal{A}$ as input (i.e., a weighted automaton in which all weights are $\bar{1}$) and adds weighted transitions to it until no more can be added. The addition of transitions is based on the following rule: for a WPDS rule $r = \langle p, \gamma \rangle \hookrightarrow \langle q, \gamma_1 \cdots \gamma_n \rangle$ with weight $f(r)$ and transitions $(q, \gamma_1, q_1), \cdots, (q_{n-1}, \gamma_n, q_n)$ with weights $w_1, \cdots, w_n$, add the transition (p, γ, q_n) to $\mathcal{A}$ with weight $w = f(r) \otimes w_1 \otimes \cdots \otimes w_n$. If this transition already exists with weight w', change the weight to $w \oplus w'$.

The algorithm is based on the intuition that if the automaton accepts configurations c and c' with weights w and w', respectively, and rule r allows the transition $c' \Rightarrow c$, then the automaton needs to accept c' with weight $w' \oplus (f(r) \otimes w)$. Termination follows from the fact that the number of states of the automaton does not increase (hence, the number of transitions is bounded), and the fact that the weight domain satisfies the ascending-chain condition (Defn. 2.2.6, item 4).

We now provide some intuition into why one needs both forwards and backwards automata. Consider the automata in Fig. 2.9(b). For the *poststar* automaton, when one follows a path that accepts the configuration $\langle p, n_8\ n_4 \rangle$, the transition (p, n_8, q) comes before (q, n_4, acc). However, the former transition describes the transformation inside `bar`, which happens *after* the transformation performed in reaching the call site at n_4 (which is stored on (q, n_4, acc)). Because the transformation for the calling context happens earlier in the program, but its transitions appear later in the automaton, the weights are read backwards. For the *prestar* automaton, the weight on (p, n_4, acc) is the transformation for going from n_4 to n_6, which occurs after the transformation inside `bar`. Thus, it is a forwards automaton.

The following lemma states the complexity for solving *poststar* by the algorithm of Reps et al. [218]. We will assume that the time to perform an extend and a combine are the same, and use the notation $O_s(.)$ to denote the time bound in terms of semiring operations. The *height* of a weight domain is defined to be the length of the longest ascending chain in the domain. For ease of stating a complexity result, we will assume that there is a finite upper bound on the height. Some weight domains, such as $\mathcal{M}$ in Defn. 2.2.10, have no such finite upper bound on the height; however, WPDSs can still be used when the height is unbounded. The absence of infinite ascending chains (Defn. 2.2.6, item 4) ensures that saturation-based algorithms for computing $post^*$ and pre^* will eventually terminate.

Lemma 2.2.14 *[218] Given a WPDS with PDS $\mathcal{P} = (P, \Gamma, \Delta)$, if $\mathcal{A} = (Q, \Gamma, \rightarrow, P, F)$ is a $\mathcal{P}$-automaton that accepts an input set of configurations, poststar produces a backward weighted automaton with at most $|Q| + |\Delta|$ states in time $O_s(|P||\Delta|(|Q_0| + |\Delta|)H + |P||\lambda_0|H)$, where $Q_0 = Q \backslash P$, $\lambda_0 \subseteq \rightarrow$ is the set of all transitions leading from states in Q_0, and H is the height of the weight domain.* □

Approximate Analysis.

Among the properties imposed by a weight domain, one important property is distributivity (Defn. 2.2.6, item 2). This is a common requirement for a precise analysis, which also arises in various *coincidence theorems* for dataflow analysis [145; 150; 240]. Sometimes this requirement is too strict and may be relaxed to monotonicity, i.e., for all $a, b, c \in D$, $(a \otimes b) \oplus (a \otimes c) \sqsubseteq a \otimes (b \oplus c)$ and $(a \otimes c) \oplus (b \otimes c) \sqsubseteq (a \oplus b) \otimes c$. In such cases, the JOVP computation may not be precise, but it will be *safe* under the partial order $\sqsubseteq$.

2.2.4.4   Local Variables and Extended Weighted Pushdown Systems

This section discusses an extension of WPDSs that permits abstractions to track the values of local variables [152].

In WPDSs, reachability problems compute the value of a rule sequence by taking an extend of the weights of each of the rules in the sequence; when WPDSs are used for dataflow analysis of a program, rule sequences represent interprocedural paths in the program. To summarize the weights of such paths, we have to maintain information about local variables of all unfinished procedures that appear on the path.

Extended WPDSs (EWPDSs) lift WPDSs to handle local variables in much the same way that Knoop and Steffen lifted conventional dataflow-analysis algorithms to handle local variables [150]: at a call site at which procedure P calls procedure Q, the local variables of P are modeled as if the current incarnations of P's locals are stored in locations that are inaccessible to Q and to procedures transitively called by Q—consequently, the contents of P's locals cannot be affected by the call to Q; we use special merging functions to combine them with the value returned by Q to create the state after Q returns.[6]

For a semiring $\mathcal{S}$ on domain D, a *merging function* is defined as follows:

Definition 2.2.15 *A function* $m : D \times D \rightarrow D$ *is a* **merging function** *with respect to a bounded idempotent semiring* $\mathcal{S} = (D, \oplus, \otimes, \overline{0}, \overline{1})$ *if it satisfies the following properties.*

1. **Strictness.** *For all* $a \in D$, $m(\overline{0}, a) = m(a, \overline{0}) = \overline{0}$.

2. **Distributivity.** *The function distributes over* $\oplus$. *For all* $a, b, c \in D$,

$$m(a \oplus b, c) = m(a, c) \oplus m(b, c) \quad and \quad m(a, b \oplus c) = m(a, b) \oplus m(a, c)$$

□

---

[6]Note that this model agrees with programming languages like Java, where it is not possible to have pointers to local variables (i.e., pointers into the stack). For languages such as C and C++, where the address-of operator (&) allows the address of a local variable to be obtained, if P passes such an address to Q, it is possible for Q (or a procedure transitively called from Q) to affect a local of P by making an indirect assignment through the address.

Conventional interprocedural dataflow-analysis algorithms must also worry about this issue, which is usually dealt with by (i) performing a preliminary analysis to determine which call sites might have such effects, and (ii) using the results of the preliminary analysis to create sound transformers for the primary analysis. The preliminary analysis is itself an interprocedural dataflow analysis, and (E)WPDSs can be applied to this problem as well. For instance, [215] discusses how one such preliminary analysis—alias analysis for single-level pointers [155]—can be expressed as a reachability problem in an EWPDS.

Definition 2.2.16 *Let $(\mathcal{P}, \mathcal{S}, f)$ be a weighted pushdown system; let $\mathcal{G}$ be the set of all merging functions on semiring $\mathcal{S}$, and let Δ_2 denote the set of push rules of $\mathcal{P}$. An **extended weighted pushdown system** is a quadruple $\mathcal{W}_e = (\mathcal{P}, \mathcal{S}, f, g)$ where $g : \Delta_2 \to \mathcal{G}$ assigns a merging function to each rule in Δ_2.* $\square$

Note that a push rule has both a weight and a merging function associated with it. Merging functions are used to fuse the local state of the calling procedure as it existed just before the call with the effects on the global state produced by the called procedure.

Consider the ICFG shown in Fig. 2.5. The rules of the PDS that represents the ICFG are listed in Section 2.2.2. We can perform constant propagation (with uninterpreted expressions) by assigning a weight to each PDS rule. The weight semiring is $\mathcal{S} = (D, \oplus, \otimes, \bar{0}, \bar{1})$, where $D = (Env \to Env)$ is the set of all environment transformers, and the semiring operations and constants are defined as follows:

$$\bar{0} = \lambda e.\bot \qquad w_1 \oplus w_2 = \lambda e.(w_1(e) \sqcup w_2(e))$$
$$\bar{1} = \lambda e.e \qquad w_1 \otimes w_2 = w_2 \circ w_1$$

The weights for the EWPDS that models the program in Fig. 2.5 are shown as edge labels. The merging function for the rule $\langle p, n_3 \rangle \hookrightarrow \langle p, e_f n_4 \rangle$, which encodes the call at n_3, receives two environment transformers: one that summarizes the effect of the caller from its enter node to the call site (e_{main} to n_3) and one that summarizes the effect of the called procedure (e_f to x_f). The merging function has to produce the transformer that summarizes the effect of the caller from its enter node to the return site (e_{main} to n_4). The merging function is defined as follows:

$$m(w_1, w_2) = \textbf{if } (w_1 = \bar{0} \text{ or } w_2 = \bar{0}) \textbf{ then } \bar{0}$$
$$\textbf{else } \lambda e.e[a \mapsto w_1(e)(a), y \mapsto (w_1 \otimes w_2)(e)(y)]$$

This copies over the value of the local variable a from the call site and gets the value of y that is returned from the called procedure. Because the merging function has access to the environment transformer just before the call, we do not have to pass the value of local variable a into procedure p. Hence the call stops tracking the value of a using the weight $\lambda e.e[a \mapsto \top, b \mapsto e(a)]$.

The merging function for the rule $\langle p, n_7 \rangle \hookrightarrow \langle p, e_f n_8 \rangle$ is defined similarly.

Merging Functions for Boolean Programs. In this section, we assume without loss of generality that each procedure has the same number of local variables.

To encode Boolean programs that have local variables, let G be the set of valuations of the global variables and L be the set of valuations of local variables. The actions of program statements and conditions are now binary relations on $G \times L$; thus, the weight domain is a relational weight domain on the set $G \times L$ but with an extra merging function defined on weights. Because different weights can refer to local variables from different procedures, one cannot take relational composition of weights from different procedures. The *project* function is used to change the scope of a weight. It existentially quantifies out the current transformation on local variables and replaces it with an identity relation. Formally, it can be defined as follows:

$$project(w) = \{(g_1, l_1, g_2, l_1) \mid (g_1, l_1, g_2, l_2) \in w\}.$$

Once the summary of a procedure is calculated as a weight w involving local variables of the procedure, the *project* function is applied to it, and the result *project*(w) is passed to the callers of that procedure. This makes sure that local variables of one procedure do not interfere with those of another procedure. Thus, merging functions for Boolean programs all have the form

$$m(a, b) = a \otimes project(b).$$

For encoding Boolean programs with other abstractions, such as finding the shortest trace, one can use the relational weight domain on $(G \times L, \mathcal{S})$, where $\mathcal{S}$ is a weight domain such as the minpath semiring (transparent to the presence or absence of local variables). The *project* function on weights from this domain can be defined as follows:

$$project(w) = \lambda(g_1, l_1, g_2, l_2). \quad \begin{array}{l} \text{if } (l_1 \neq l_2) \text{ then } \overline{0}_{\mathcal{S}} \\ \text{else } \bigoplus_{l \in L} w(g_1, l_1, g_2, l) \end{array}$$

Again, the merging functions all have the form $m(a, b) = a \otimes project(b)$.

2.2.4.5 Polyhedral Analysis with WPDSs

Recently, Gopan [118] presented a way to perform numeric program analysis with WPDSs using the polyhedral abstract domain [79]. One of the challenges that he faced was that the polyhedral domain has infinite ascending chains, and hence widening techniques are required [77].

Widening is implemented using a weight wrapper that supports the normal weight interface extended with a few extra methods. Two types of weights are used: "regular weights" and "widening weights". Regular weights behave just like ordinary weights; widening weights are placed on WPDS rules where widening must occur (e.g., rules that correspond to backedges in the ICFG). In particular, if a widening weight b is used in a combine operation by the WPDS saturation procedure, the normal operation $a \oplus b$ is replaced by $a \triangledown (a \oplus b)$, (where $\triangledown$ is the standard widening operator).

2.3   DATALOG

Datalog is a subset of first-order logic. It is also a restricted form of the logic programming language Prolog. At the same time, Datalog is also a query language for databases. In information security, Datalog has been used widely as a tool for knowledge representation and analysis.

Before giving the syntax and semantics of Datalog, we first look at an example Datalog program. We follow the convention of Prolog, where an identifier starting with an upper-case letter represents a variable and an identifier starting with a lower-case letter represents a predicate symbol or a constant symbol.

Example 2.1   The following Datalog program consists of four clauses.

$$in\_role(alice, accountant).$$
$$is\_senior(accountant, clerk).$$
$$is\_senior(clerk, employee).$$
$$in\_role(X, R1) \quad \leftarrow \quad in\_role(X, R2), is\_senior(R2, R1).$$

The four datalog clauses above represent the following pieces of knowledge, respectively: (1) "Alice is member of the accountant role", (2) "the accountant role is senior to the clerk role", (3) "the clerk role is senior to the employee role", and (4) "members of one role are also members of any role that the role is senior to".

In the above program, *in_role* and *is_senior* are predicate symbols, *alice*, *accountant*, *clerk*, and *employee* are constant symbols, and X, Y, Z are variables. Intuitively, from the above program, we should be able infer that *in_role(alice, clerk)* and *in_role(alice, employee)*.

To formally define the syntax of datalog, we now review some standard terminology of first-order logic (FOL). An alphabet $\mathcal{A}$ is given by a set of *variables*, a set of *function symbols*, and a set of *predicate symbols*. Each function symbol and each predicate symbol has an associated arity, which is a non-negative integer. A function symbol that has arity 0 is also called a *constant*.

The set of *terms* of a given alphabet $\mathcal{A}$ is the smallest set $\mathcal{T}$ such that:

1. Any constant in $\mathcal{A}$ is in $\mathcal{T}$;

2. Any variable in $\mathcal{A}$ is in $\mathcal{T}$;

3. If f is a function symbol in $\mathcal{A}$ with arity $n \geq 1$ and $t_1, \ldots, t_n \in \mathcal{T}$, then $f(t_1, \ldots, t_n) \in \mathcal{T}$.

An *atomic formula* (or simply an *atom*) takes the form $p(t_1, \ldots, t_k)$, where p is a predicate symbol of arity k and $t_1, \ldots, t_k$ are terms. In FOL, formulas are formed using atoms, *logical connectives* $\wedge$ (conjunction), $\vee$ (disjunction), $\neg$ (negation), $\rightarrow$ (implication), and $\leftrightarrow$ (logical equivalence), and *quantifiers* $\forall$ (universal) and $\exists$ (existential).

In FOL, a *literal* is either an atom (also called a *positive literal*) or the negation of an atom (also called a *negative literal*), and a *clause* is a disjunction of literals. An FOL clause takes the following form

$$\forall_{X_1} \forall_{X_2} \cdots \forall_{X_m} (L_0 \vee L_1 \vee \cdots \vee L_n),$$

where each L_i is a literal, and $X_1, X_2, \ldots, X_m$ are all the variables that appear in $(L_0 \vee L_1 \vee \cdots \vee L_n)$. Because all quantifiers in a clause are universal quantifiers, we usually omit them when writing a clause.

Logic programming in general, and Datalog in particular, uses Horn clauses. A *Horn clause* is a clause with *at most one* positive literal. A clause with *exactly one* positive literal is called a *definite clause*, and a clause with *zero* positive literal is called a *goal clause*. That is, a Horn clause is either a definite clause or a goal clause.

A definite clause is of the form (with quantifiers omitted)

$$A_0 \vee \neg A_1 \vee \neg A_2 \vee \cdots \vee \neg A_n$$

where each A_i is an atom, i.e., it takes the form $p(t_1, \ldots, t_k)$ such that p is a *predicate symbol* and $t_1, \ldots, t_k$ are *terms*. The definite clause above can be equivalently written as

$$A_0 \vee \neg (A_1 \wedge A_2 \wedge \cdots \wedge A_n),$$

and as

$$(A_1 \wedge A_2 \wedge \cdots \wedge A_n) \rightarrow A_0,$$

in which the "$\rightarrow$" symbol denotes logical implication. The notational convention is to write such a definite clause as:

$$A_0 \leftarrow A_1, \ldots, A_n$$

In the logic programming language Prolog, the symbol :− is often used to replace $\leftarrow$.

In the above clause, the left-hand side (i.e., A_0) is called the *head* and the right-hand side (i.e., $A_1, \ldots, A_n$) is called the *body*. The body of a clause may be empty, in which case the symbol $\leftarrow$ is often omitted. Intuitively, such a clause means that if all atoms in the body are true then the atom in the head is also true. A clause with an empty body is called a *fact*, as intuitively it means that the literal in the head is always true. A clause with a nonempty body is called a *rule*.

A Horn clause is a Datalog clause if it does not have function symbols that have arity greater than zero. That is, a term in Datalog is either a constant or a variable. A *Datalog program* consists of a collection of Datalog definite clauses.

In information security, a Datalog program can encode security state information such as a set of access-control policies or the firewall configurations of a network. Using facts, one can describe elements that are known to be true in the security state. Using rules, one can describe the logical implication relationships in the security state, i.e., what must be true if some preconditions are known to be true. One can then issue analysis queries against the Datalog program, verifying whether certain properties hold or not in the encoded state.

As Datalog is a subset of FOL, one can view a Datalog program P as an FOL theory and express a query using an FOL formula F. The query F is true given P if and only if the formula F is a *logical consequence* of P, written as $P \models F$. In model theory, $P \models F$ is true if and only if F is true in every model of P. A nice property of Datalog is that when we limit the query formula F to be the negation of a goal clause, the notion of logical consequence can be defined in a simple way using the least Herbrand model of the program P, and query evaluation can be performed efficiently.

To define the least Herbrand model of a Datalog program, we now give the definitions for the more general case of logic programs, which may contain function symbols of arity greater than zero.

Definition 2.2   Let $\mathcal{A}$ be an alphabet containing at least one constant. A literal, fact, rule, or clause that does not contain any variable is said to be *ground*. The set $U_{\mathcal{A}}$ of all ground terms constructed from function symbols and constants in $\mathcal{A}$ is called the *Herbrand Universe* of $\mathcal{A}$. The set of all ground, atomic formulas over $\mathcal{A}$ is called the *Herbrand base* of $\mathcal{A}$. A Herbrand interpretation I of a logic program P is a subset of the Herbrand base of P.

The Herbrand universe and Herbrand base is often defined over a given program. In this case, it is assumed that the alphabet of the program consists of exactly those symbols that appear in the program. Given a logic program P, a Herbrand interpretation I assigns a truth value to each atom in the Herbrand base of P: a ground atom a is true if and only if $a \in I$.

Definition 2.3   A ground rule $a_0 \leftarrow a_1, \ldots, a_n$ is satisfied by a Herbrand interpretation I if either $a_0 \in I$ or at least one of $a_1, \cdots, a_n$ is not in I. We say that a clause $A_0 \leftarrow A_1, \ldots, A_n$ is satisfied by I if every ground instantiation of the clause (i.e., substituting the variables in the clause with terms in the Herbrand universe of P consistently) is satisfied by I. We say that I is a *Herbrand model* of P if each clause in P is satisfied by I.

Some important results about Herbrand models of logic programs are as follows. Given any two Herbrand models of a program, the intersection of the two models is also a model of the program. A definite clause is always positive; in that, it specifies only what must hold and not what must not hold. As a result, each program must have at least one model. For example, the Herbrand base (the maximum interpretation) is always a model. It follows that for any logic program, there must exist a unique least Herbrand Model. Finally, a ground fact logically follows from a program if and only if it is in the least Herbrand model.

The least Herbrand model of the program in Example 2.1 is

$\{$ *in_role(alice, accountant)*,   *is_senior(accountant, clerk)*,   *in_role(alice, clerk)*,
    is_senior(clerk, employee),   *in_role(alice, employee)* $\}$.

We now summarize a standard fixpoint characterization of the least Herbrand model. Note that for a datalog program both the Herbrand Universe and the Herbrand base are finite. Given a datalog program P, let P^{inst} be the ground instantiation of P using constants in P; the *immediate consequence operator*, T_P, is defined as follows: Given a set K of ground logical atoms, $T_P(K)$ consists of all logical atoms, a, such that there is a ground clause $a \leftarrow b_1, \ldots, b_n$ in P^{inst}, where $n \geq 0$, and either $n = 0$ or $b_j \in K$ for $1 \leq j \leq n$. The least fixpoint of T_P is given by

$$T_P{\uparrow}^{\omega} = \bigcup_{i=0}^{\infty} T_P{\uparrow}^i, \ \text{where } T_P{\uparrow}^0 = \emptyset \text{ and } T_P{\uparrow}^{i+1} = T_P(T_P{\uparrow}^i), i \geq 0$$

$T_P{\uparrow}^1$ includes all facts in P, $T_P{\uparrow}^2$ includes all facts in P as well as all atoms that are logically implied by facts in P, $T_P{\uparrow}^3$ includes all atoms in $T_P{\uparrow}^2$ as well as all atoms that are logically implied by atoms in $T_P{\uparrow}^2$. The sequence $T_P{\uparrow}^i$ is a non-decreasing sequence of subsets of a finite set. Thus, there exists an N such that $T_P(T_P{\uparrow}^N) = T_P{\uparrow}^N$ and $T_P{\uparrow}^{\omega} = T_P{\uparrow}^N$. $T_P{\uparrow}^{\omega}$ is identical to the least Herbrand model of P [167].

It has been shown that the least Herbrand model of a ground program can be computed in time linear in its size [99]. Thus the least Herbrand model of P can be computed in time linear in the size of P^{inst}. If the total size of P is M, then there are no more than M constants in P. Assuming that the number of variables in each clause is bounded by a constant, v, the number of instances of each clause is therefore $O(M^v)$, so the size of P^{inst} is $O(M^{v+1})$.

The computation of the least Herbrand model of a datalog program has been studied in the deductive-database literature. In the deductive-database setting, the number of facts is often large and the number of rules is often small: the facts correspond to tuples in input tables, and the rules together correspond to a query. The fixpoint-based computation of the least Herbrand model is known as the naive *bottom-up* evaluation algorithm. The seminaive evaluation algorithm focuses on the new facts generated in each iteration and avoids recomputing the same facts [13].

Given a Datalog program P, the simplest query is a single ground atom a and asks whether the atom is true given P. This can be answered by computing the least Herbrand model of P and checking whether a is in the model. Datalog allows more sophisticated forms of queries. A Datalog query can be the negation of a goal clause. Recall that a goal clause is a clause that has only negative literals. It takes the form:

$$\forall X_1 \cdots \forall X_m \, (\neg A_1 \vee \cdots \vee \neg A_n)$$

where $X_1, \ldots, X_m$ are the variables in $A_1, \ldots, A_n$. This is equivalent to

$$\forall X_1 \cdots \forall X_m \neg (A_1 \wedge \cdots \wedge A_n) .$$

The notational convention is to write it as

$$\leftarrow A_1, \ldots, A_n$$

The above goal clause is logically equivalent to

$$\neg \exists X_1 \cdots \exists X_m \, (A_1 \wedge \cdots \wedge A_n) .$$

The query intended by the above clause is the negation of the above clause, i.e.,

$$F = \exists X_1 \cdots \exists X_m \, (A_1 \wedge \cdots \wedge A_n) .$$

For example, given the logic program in Example 2.1, the query "$\exists X (in\_role(X, accountant) \wedge in\_role(X, clerk))$" asks whether there exists a user who is a member of both the accountant role and the clerk role. The answer to this query is yes, as $(in\_role(alice, accountant) \wedge in\_role(alice, clerk))$ is a logical consequence of the program.

Given a program P and a query $F = \exists X_1 \cdots \exists X_m \, (A_1 \wedge \cdots \wedge A_n)$, to check whether $P \models F$, one can use the resolution principle. That is, one consider whether the FOL theory $P \cup \neg F$ has a model; $P \models \neg F$ holds if and only if $P \cup \neg F$ *does not have* a model. Note that $\neg F$ is the goal clause "$\leftarrow A_1, \ldots, A_n$". The basic inference rule used in logic programming is SLD resolution (Selective Linear Definite clause resolution). Using SLD resolution to determine whether $P \models F$ holds is known as *top-down* inference, or goal-directed inference, as the inference starts from the goal clause. The standard SLD resolution may not terminate, because it may revisit the same resolution goal again and again. The logic programming system XSB [121] implements SLG resolution, which enhances SLD resolution with tabling. The SLG resolution is guaranteed to terminate for datalog.

The XSB logic programming system has been used extensively when datalog is used for analysis in security.

CHAPTER 3

Detecting Buffer Overruns Using Static Analysis

Buffer overruns are one of the most exploited class of security vulnerabilities. In a study by the SANS institute [5], buffer overruns in RPC services ranked as the top vulnerability in UNIX systems. A simple mistake on the part of a careless programmer can cause a serious security problem. Consequences can be as serious as a remote user acquiring root privileges on the vulnerable machine. To add to the problem, these vulnerabilities are easy to exploit, and several "cookbooks" [7; 242] are available to construct such exploits. As observed by several researchers [156; 253], C is highly vulnerable because there are several library functions that manipulate buffers in an unsafe way. Millions of lines of legacy code have been written in C, and systems running these applications continue to be vulnerable.

Several approaches have been proposed to mitigate the problem – these range from dynamic techniques [45; 65; 73; 80; 200; 211] that *prevent* attacks based on buffer overruns, to static techniques [98; 156; 222; 252; 253] that examine source code to *eliminate* these bugs before the code is deployed. Combinations of static and dynamic techniques have also been proposed in which the results of static analysis are used to remove run-time checks. Unlike static techniques, dynamic techniques do not eliminate bugs, and often have the undesirable effect of causing the application to crash when an attack is discovered. Static techniques have the added advantage that they impose no run-time overhead on applications.

In this chapter, we describe the design and implementation of a tool that statically analyzes C source code to detect buffer-overrun vulnerabilities.[1] This chapter demonstrates the following points:

- The use of static analysis to model C string manipulations as a linear program.

- The design and implementation of fast, scalable solvers based on novel use of techniques from the linear-programming literature. The solution to the linear program determines buffer bounds.

- Techniques to make the program analysis context-sensitive.

- The efficacy of other program-analysis techniques, such as static slicing to understand and eliminate bugs from source code.

---

[1]This tool was developed in a collaboration between the University of Wisconsin and Grammatech, Inc.

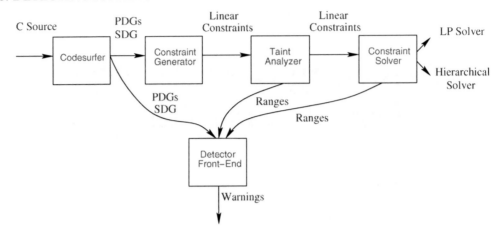

Figure 3.1: Overall Architecture of the Buffer-Overrun Tool.

One of our principal design goals was to make the tool scale to large real world applications. We used the tool to audit several popular and commercially used packages. The tool identified 14 previously unknown buffer overruns in wu-ftpd-2.6.2 (Section 3.5.1.1) in addition to several known vulnerabilities in other applications.

The rest of the chapter is organized as follows: Section 3.1 describes the overall architecture of the tool. Section 3.2 and Section 3.3 describe the design of two solvers used by the tool. Section 3.4 describes a technique to make the program analysis context-sensitive. We report on our experience with the prototype implementation in Section 3.5. Section 3.6 discusses other work on buffer-overrun detection.

3.1   OVERALL TOOL ARCHITECTURE

The tool has five components (Figure 3.1) that are described in the remainder of this section. Section 3.1.1 describes the code-understanding tool CodeSurfer. CodeSurfer is used by the *constraint generator*, the *detector front-end*, and to help the user examine potential overruns. Section 3.1.2 describes constraint generation. Section 3.1.3 presents *taint analysis*, which identifies and removes unconstrained constraint variables. Section 3.1.4 describes how to the constraint system is solved. Section 3.1.5 explains how the solution to the constraint system is used to detect potential buffer overruns. The program in Figure 3.2 will serve as a running example.

3.1.1   CODESURFER

The constraint generator and the detector front-end are both developed as plug-ins to CodeSurfer. CodeSurfer is a code-understanding tool that was originally designed to compute precise interprocedural slices [135; 136]. CodeSurfer builds a whole program representation that includes a system

```
(1)  main(int argc, char* argv[]){
(2)      char header[2048], buf[1024],
             *cc1, *cc2, *ptr;
(3)      int counter;
(4)      FILE *fp;
(5)      ...
(6)      ptr = fgets(header, 2048, fp);
(7)      cc1 = copy_buffer(header);
(8)      for (counter = 0; counter < 10;
                            counter++){
(9)          ptr = fgets (buf, 1024, fp);
(10)         cc2 = copy_buffer(buf);
(12)     }
(13) }
(14)
(15) char *copy_buffer(char *buffer){
(16)     char *copy;
(17)     copy = (char *) malloc(strlen(buffer));
(18)     strcpy(copy, buffer);
(19)     return copy;
(20) }
```

Figure 3.2: Running Example.

dependence graph (which is composed of program dependence graphs for each procedure), an interprocedural control-flow graph, abstract syntax trees (ASTs) for program expressions, side-effect information, and points-to information. CodeSurfer presents the user with a GUI for exploring these program representations. The queries that CodeSurfer supports include forward and backward slicing from a program point, precise interprocedural chopping between two program points (for details, see [216]), finding data and control dependence predecessors and successors from a program point, and examining the points-to set of a program variable. CodeSurfer presents the user with a listing of their source code that is "hot", i.e., the user can click on a program point in the code and ask any of the queries listed above.

CodeSurfer has two primary uses in the buffer-overrun tool: (1) the constraint generator is a CodeSurfer plug-in that makes use of CodeSurfer's ASTs and pointer analysis (an implementation of Andersen's analysis [21]). (2) the detector front-end is a CodeSurfer plug-in that uses CodeSurfer's GUI to display potential overruns. Information about potential overruns is linked to CodeSurfer's internal program representation so that the user can make use of CodeSurfer's features, such as slicing, to examine potential overruns.

3.1.2   CONSTRAINT GENERATION

Constraint generation is similar to the approaches proposed in [98; 156; 252]. We also use points-to information returned by Codesurfer, thus allowing more precise constraints to be generated. Each pointer `buf`, of type pointer to character buffer, is modeled by four constraint variables – `buf!used!max` and `buf!used!min`, which denote the maximum and minimum number of bytes used in the buffer, and `buf!alloc!max` and `buf!alloc!min`, which denote the maximum and minimum number of bytes allocated for the buffer.

Constraint	Program Statement
`header!used!max ≥ 2048`	6
`header!used!min ≤ 1`	6
`buffer!used!max ≥ buf!used!max`	10 (function call)
`buffer!used!min ≤ buf!used!min`	10 (function call)
`buffer!alloc!max ≥ buf!alloc!max`	10 (function call)
`buffer!alloc!min ≤ buf!alloc!min`	10 (function call)
`copy_buffer$return!alloc!max ≥ copy!alloc!max`	19
`copy_buffer$return!alloc!min ≤ copy!alloc!min`	19
`copy_buffer$return!used!max ≥ copy!used!max`	19
`copy_buffer$return!used!min ≥ copy!used!min`	19
`cc2!used!max ≥ copy_buffer$return!used!max`	10 (assignment)
`cc2!used!min ≤ copy_buffer$return!used!min`	10 (assignment)
`cc2!alloc!max ≥ copy_buffer$return!alloc!max`	10 (assignment)
`cc2!alloc!min ≥ copy_buffer$return!alloc!min`	10 (assignment)
`counter'!max ≥ counter!max + 1`	8 (counter++)
`counter!max ≥ counter'!max`	8 (counter++)
`counter'!min ≤ counter!min + 1`	8 (counter++)
`counter!min ≤ counter'!min`	8 (counter++)

Figure 3.3: Some constraints for the running example.

Each integer variable `i` is modeled by the constraint variables `i!max` and `i!min`, which represent the maximum and minimum value of `i`, respectively. Program statements that operate on pointers to character buffers or on integer variables are modeled using linear constraints over constraint variables.

Our constraints model the program in a *flow-* and *context-insensitive* manner, with the exception of library functions that manipulate pointers to character buffers (also known as string buffers). A flow-insensitive analysis ignores the order of statements, and a context-insensitive analysis does not differentiate between multiple call-sites to the same function. For a function call to a library function that manipulates a buffer string (e.g., `strcpy` or `strlen`), we generate constraints that model the effect of the call; for these functions, the constraint model is context-sensitive. In Section 3.4, we will show how we extended the model to make the constraints context-sensitive for user-defined functions as well.

Constraints are generated using a single pass over the program's statements. There are four program statements for which constraints are generated: buffer declarations, assignments, function calls, and return statements. A buffer declaration such as `char buf[1024]` results in constraints that indicate that `buf` is of size 1024. A statement that assigns into a string buffer (e.g., `buf[i]='c'`) results in constraints that reflect the effect of the assignment on `buf!used!max` and `buf!used!min`. An assignment to an integer `i` results in constraints on `i!max` and `i!min`.

As mentioned above, a function call to a library function that manipulates a string buffer is modeled by constraints that summarize the effect of the call. For example, the `strcpy` statement at line (18) in Figure 3.2 results in the following constraints:

$$copy!used!max \geq buffer!used!max$$
$$copy!used!min \leq buffer!used!min$$

For each user-defined function `foo`, there are constraint variables for `foo`'s formal parameters that are integers or string buffers. If `foo` returns an integer or a string buffer, then there are constraint variables (e.g., `copy_buffer$return!used!max`) for the function's return value. A call to a user-defined function is modeled with constraints for the passing of actual parameters and the assignment of the function's return value.

As in [252], constraints are associated with pointers to string buffers rather than the string buffers themselves. This means that some aliasing among string buffers is not modeled in the constraints, and false negatives may result. We chose to follow [252] in this regard because we are interested in improving precision by using a context-sensitive program analysis (Section 3.4). Currently, context-sensitive pointer analysis does not scale well, and using a context-insensitive pointer analysis would undermine our objective of performing context-sensitive buffer-overrun analysis.

However, we discovered that we could make use of pointer analysis to eliminate some false negatives. For instance, consider the statement "`strcpy(p->f, buf)`," where p could point to a structure s. The constraints generated for this statement would relate the constraint variables for `s.f` and `buf`. Moreover, we use the results of pointer analysis to handle arbitrary levels of dereferencing. Constraint generation also makes use of pointer information for integer pointers.

Input: Set of Constraints C
Output: Subset of C with no uninitialized, or infinite variables
(1)   InfSet = {var | var $\leq -\infty \vee \in C$ var $\geq \infty \in C$} $\cup$ {var | var is uninitialized}
(2)   **while** InfSet $\neq \phi$
(3)   Select and remove var from InfSet
(4)   **foreach** Constraint $c \in C$ of the form MaxVar $\geq$ RHS
(5)   **if** MaxVar is var
(6)   Drop c from C
(7)   **else if** var appears in RHS
(8)   Set MaxVar to $+\infty$ and add MaxVar to InfSet
(9)   Drop c from C
(10)   **endif**
(11)   **foreach** Constraint $c \in C$ of the form MinVar $\leq$ RHS
(12)   **if** MinVar is var
(13)   Drop c from C
(14)   **else if** var appears in RHS
(15)   Set MinVar to $-\infty$ and add MinVar to InfSet
(16)   Drop c from C
(17)   **endif**
(18)   Return C

Figure 3.4: Algorithm for Taint Analysis.

Figure 3.3 shows a few constraints for the program in Figure 3.2, along with the program statement that generated them. Most of the constraints are self-explanatory, however a few comments are in order:
• Because we do not model control flow, we ignore branch conditions during constraint generation. Hence we do not model the effect of the condition in an `if` or `for` statement; the condition `counter < 10` in line (8) was ignored in our example.
• The statement `counter++` is particularly interesting when generating linear constraints. A linear constraint such as `counter!max` $\geq$ `counter!max + 1` cannot be interpreted by a linear program solver.

Hence, we model this statement by treating it as a pair of statements: counter' = counter + 1; counter = counter'. These two constraints capture the fact that counter has been incremented by 1, and can be translated into constraints that are acceptable to a linear program solver, although the resulting linear program will be *infeasible*. Section 3.2 discusses these and related issues in detail.

• A program variable that acquires its value from the environment or from user input in an unguarded manner is considered unsafe – for instance, the statement getenv("PATH"), which returns the search path, could return an arbitrarily long string. To reflect the fact that the string can be arbitrarily long, we generate constraints getenv\$return!used!max $\geq \infty$ and getenv\$return!used!min ≤ 0. Similarly, an integer variable i accepted as user input gives rise to constraints i!max $\geq \infty$ and i!min $\leq -\infty$.

3.1.3   TAINT ANALYSIS

The linear constraints then pass through a *taint-analysis* module. In Sections 3.2 and 3.3, we will demonstrate two techniques to solve the constraints using linear programming. The main goal of the taint-analysis module is to make the constraints amenable to these solvers. Linear programming can work only with finite values, hence this requires us to remove variables that can obtain infinite values. Moreover, it is also important that max variables have finite lower bounds and min variables have finite upper bounds. Hence, the objectives of this module are two-fold:

• *Identify and remove any variables that get an infinite value*: As mentioned in section 3.1.2, some constraint variables var are associated with constraints of the form var $\geq \infty$ or var $\leq -\infty$. Taint analysis identifies constraint variables that can directly or indirectly be set to $\pm\infty$ through such constraints and removes them from the set of constraints.

• *Identify and remove any uninitialized constraint variables*: The system of constraints is examined to see if all max constraint variables have a finite lower bound, and all min constraint variables have a finite upper bound; we refer to constraint variables that do not satisfy this requirement as *uninitialized*. Constraint variables may fail to satisfy the above requirement if either the program variables that they correspond to have not been initialized in the source code, or program statements that affect the value of the program variables have not been captured by the constraint generator. The latter case may arise when the constraint generator does not have a model for a library function that affects the value of the program variable. It is important to realize that this analysis is not meant to capture uninitialized *program* variables, but is meant to capture uninitialized *constraint* variables.

Figure 3.4 presents the taint-analysis algorithm. In the constraints obtained by the program in Figure 3.2, no variables will be removed by the taint analysis module, assuming that we modeled the library functions strlen, fgets, and strcpy correctly.

3.1.4   CONSTRAINT SOLVING

The constraints that remain after taint analysis can be solved using linear programming. We have developed two solvers, both of which use linear programming to obtain values for the constraint variables. The first method uses a linear program solver on the entire set of constraints to obtain values for constraint variables; a detailed description of the algorithm can be found in Section 3.2.

The second method analyzes and breaks up the set of constraints into smaller subsets, and passes each of these subsets to the linear program solver; we explain this algorithm in Section 3.3.

The goal of both solvers is the same, to obtain the best possible estimate of the number of bytes used and allocated for each buffer in any execution of the program. For a buffer pointed to by buf, finding the number of bytes used corresponds to finding the "tightest" possible range [buf!used!min..buf!used!max]. This can be done by finding the lowest and highest values of the constraint variables buf!used!max and buf!used!min, respectively, that satisfy all the constraints. Similarly, we can find the "tightest" possible range for the number of bytes allocated for the buffer by finding the lowest and the highest values of buf!alloc!max and buf!alloc!min, respectively.

For the program in Figure 3.2, the constraint variables take on the values shown in Figure 3.5. We explain in detail in Sections 3.2 and 3.3 how these values were obtained.

Variable	min Value	max Value
header!used	1	2048
header!alloc	2048	2048
buf!used	1	1024
buf!alloc	1024	1024
cc1!used	0	2048
cc1!alloc	0	2047
ptr!used	1	2048
ptr!alloc	1024	2048
cc2!used	0	2048
cc2!alloc	0	2047
buffer!used	1	2048
buffer!alloc	1024	2048
copy!used	0	2048
copy!alloc	0	2047
counter	0	∞

Figure 3.5: Values of some constraint variables.

3.1.5   DETECTING OVERRUNS

Based on the values inferred by the solver, as well as the values inferred by the taint-analysis module, the detector decides whether there was an overrun on each buffer. We use several heuristics to give the best possible judgment. We shall explain some of these in the context of the values from Figure 3.5.

• The solver found that the buffer pointed to by header has 2048 bytes allocated for it, but that its length could have been between 1 and 2048 bytes. This is a scenario where a buffer overrun can never occur – and hence the buffer pointed to by header is flagged as "safe". The same is true of the buffer pointed to by buf.

• The buffer pointed to by ptr was found to have between 1024 and 2048 bytes allocated, while between 1 and 2048 bytes could have been used. Note that ptr is part of two assignment statements. Assignment statement (6) could make ptr point to a buffer as long as 2048 bytes, while assignment statement (9) could make ptr point to a buffer as long as 1024 bytes. The flow insensitivity of the analysis means that we do not differentiate between these program points and hence can only infer that `ptr` was up to 2048 bytes long. In such a scenario, where the value of `ptr!used!max` is bigger than `ptr!alloc!min` but smaller than (or equal to) the value of `ptr!alloc!max`, we conservatively conclude that there might have been an overrun. This can result in a *false positive* due to the flow insensitivity of the analysis.

• In cases such as for program variable `copy`, where we observe that `copy!alloc!max` is less than `copy!used!max`, we know that there is a run of the program in which more bytes are written into the buffer than it could possibly hold, and we conclude that there is a real overrun on the buffer.

Notice that the constraint variables corresponding to `cc1` and `cc2` get the same value; this is a result of the context-insensitivity of our analysis. We will show in Section 3.4 how to enhance the precision of the analysis using context sensitivity.

We have developed a GUI front end (Figure 3.14) that enables the end-user to "surf" the warnings – every warning is linked back to the source code line that it refers to. Moreover, the user can exploit the program slicing capabilities of Codesurfer to verify real overruns.

3.2 CONSTRAINT RESOLUTION USING LINEAR PROGRAMMING

A *linear program* is an optimization problem that is expressed as follows:

$$\text{Minimize} \ : \ cx$$
$$\text{Subject To} \ : \ Ax \geq b$$

where A is an $m \times n$ matrix of constants, b and c are vectors of constants, and x is a vector of variables. This is equivalent to saying that we have a system of m inequalities in n variables and are required to find values for the variables such that all the constraints in the system are satisfied and the *objective function* cx takes its lowest possible value. It is important to note that the above form is just one of several ways in which a linear program can be expressed. For a more comprehensive view of linear programming, see [68; 237]. Linear programming works on finite real numbers; that is, the variables in the vector x are only allowed to take finite real values. Hence the optimum value of the objective function, if it exists, is always guaranteed to be finite.

Linear programming is well studied in the literature, and there are well-known techniques to solve linear programs, the simplex method [85] being the most popular of them. Other known techniques, such interior-point methods [256], work in polynomial time. Commercially available solvers for solving linear programs, such as SoPlex [257; 258] and CPLEX [205], implement these and related methods.

The set of constraints that we obtained after program analysis are linear constraints, hence we can formulate our problem as a linear program. Our goal is to obtain the values for `buf!alloc!min`, `buf!alloc!max`, `buf!used!min`, and `buf!used!max` that yield the tightest possible ranges for the number of bytes allocated and used by the buffer pointed to by `buf` in such a way that all the constraints are satisfied. More precisely, we are interested in finding the lowest possible values of `buf!alloc!max` and `buf!used!max`, and the highest possible values of `buf!alloc!min` and `buf!used!min` subject to the set of constraints. We can obtain the desired bounds for each buffer `buf` by solving four linear programs, each with the same constraints but with different objective functions:

Minimize: `buf!alloc!max`
Maximize: `buf!alloc!min`
Minimize: `buf!used!max`
Maximize: `buf!used!min`

However, it can be shown (the proof is beyond the scope of this chapter) that for the kind of constraints generated by the tool, if all `max` variables have finite lower bounds, and all `min` variables have finite upper bounds, then the values obtained by solving the four linear programs as above are also the values that optimize the linear program with the same set of constraints subject to the objective function:

Minimize: $\sum_{\text{buf}}$ (`buf!alloc!max` - `buf!alloc!min`
+ `buf!used!max` - `buf!used!min`)

Note that this objective function combines the constraint variables across *all* buffers. Because taint analysis ensures that all `max` variables have finite lower bounds and all `min` variables have finite upper bounds, we can solve just *one* linear program, and obtain the bounds for *all* buffers.

It must be noted that we are actually interested in obtaining integer values that represent buffer bounds `buf!alloc!max`, `buf!used!max`, `buf!alloc!min` and `buf!used!min`. The problem of finding integer solutions to a linear program is called Integer Linear Programming and is a well known NP-complete problem [112]. Our approach is thus an approximation to the real problem of finding *integer* solutions that satisfy the constraints. In some cases, however, it is possible to solve the problem using standard linear programming algorithms and yet obtain integer solutions to the variables in the linear program. This is possible when the constraints can be expressed as $A \cdot x \geq b$, and A is a *unimodular* matrix [17; 134; 237; 250]. Here A is an $m \times n$ matrix of integer constants, x is an $n \times 1$ vector of variables, and b is an $m \times 1$ vector of integer constants. In our experience, the constraints produced by the tool have always produced integer solutions.

3.2.1 HANDLING INFEASIBLE LINEAR PROGRAMS

While at first glance, the method seems to give the desired buffer bounds, it does not work for all cases. In particular, an optimal solution to a linear program need not even exist. We describe briefly the problems faced when using a linear-programming-based approach for determining bounds on buffers.

A linear program is said to be *feasible* if one can find finite values for all the variables such that all the constraints are satisfied. For a linear program in n variables, such an assignment is a vector in $\mathbb{R}^n$ and is called a solution to the linear program. A solution is said to be *optimal* if it also maximizes (or minimizes) the value of the objective function. A linear program is said to be *unbounded* if a solution exists, but no solution optimizes the objective function. For instance, consider:

$$\text{Maximize} \ : \ \text{x}$$
$$\text{Subject To} \ : \ \text{x} \geq 5$$

Any value of $\text{x} \geq 5$ is a solution to the above linear program, but no finite value $\text{x} \in \mathbb{R}$ optimizes the objective function. Finally, a linear program is said to be *infeasible* if it has no solutions. An example of an infeasible linear program is shown in Figure 3.6.

$$\text{Minimize} \ : \ \texttt{counter!max}$$
$$\text{Subject To} \ : \ \texttt{counter'!max} \geq \texttt{counter!max} + 1$$
$$\texttt{counter!max} \geq \texttt{counter'!max}$$

Figure 3.6: An Infeasible Linear Program.

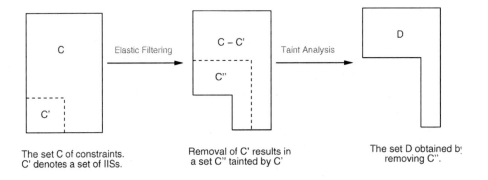

Figure 3.7: Making an infeasible set of constraints amenable to linear programming.

In our formulation, if a linear program has an optimal solution, we can use that value as the buffer bound. None of the linear programs in our case can be unbounded because the constraints have been examined by the taint analyzer to ensure that all max variables have finite lower bounds. We minimize for the max variables in the objective function, and because all the max variables have finite lower bounds, the lowest value that each max variable can obtain is also finite. Similarly, all min variables have finite upper bounds, and so when we maximize the min variables, the highest values that they could obtain are also finite. Hence taint analysis is an essential step to ensure that our approach works correctly.

However, when the linear program is infeasible, we cannot assign any finite values to the variables to obtain a feasible solution. As a result, we cannot obtain the values for the buffer bounds. In such a case, a safe option would be to set all max variables to ∞ and min variables to $-\infty$, but that information would be virtually useless to the user of the tool because there would be too many false positives. The linear program may be infeasible due to a small subset of constraints; in such a scenario,

Input: Set of Constraints C
Output: For each buffer buf, values for buf!used!max, buf!used!min, buf!alloc!max, buf!alloc!min
(1)   Removed = ϕ
(2)   **while** C is infeasible
(3)   IIS_set = $\text{E}_{\text{LASTIC}}\_\text{F}_{\text{ILTER}}\_\text{A}_{\text{LGORITHM}}(C)$
(4) $C = C - \text{IIS\_Set}$
(5)   Removed = Removed $\cup$ IIS_Set
(6)   **foreach** constraint variable v appearing in Removed
(7)   **if** v is a **max** variable
(8) $v \leftarrow \infty$
(9)   **else**
(10) $v \leftarrow -\infty$
(11) C = output of steps (2)-(12) of *Taint Analysis* (Figure 3.4) by setting Infset = Removed
(12)   MaxSet = $\{v \mid v$ is a **max** constraint variable appearing in $C\}$
(13)   MinSet = $\{u \mid u$ is a **min** constraint variable appearing in $C\}$
(14)   Minimize: $(\sum_{v \in \text{MaxSet}} v) - (\sum_{u \in \text{MinSet}} u)$ Subject To: C
(15)   Set each variable to the value returned by the Linear Program Solver.

Figure 3.8: Constraint Resolution using Linear Programming.

setting all variables to infinite values is overly conservative. For instance, the constraints in Figure 3.2 are infeasible because of the constraints generated for the statement counter++. Constraints generated by most real-world programs have such statements, as well as statements involving pointer arithmetic, and we can expect the constraints for such programs to be infeasible. Thus, the conservative approach of setting all constraint variables to infinite values is unacceptable.

We have developed an approach in which we try to remove a "small" subset of the original set of constraints so that the resultant constraint system is feasible. In fact, the problem of "correcting" infeasible linear programs to make them feasible is a well-studied problem in the operations-research community. The approach is to identify *Irreducibly Inconsistent Sets* (called *IIS*) [64]. An IIS is a minimal set of inconsistent constraints, i.e., the constraints in the IIS together are infeasible, but any subset of constraints in the IIS form a feasible set. For instance, both the constraints in the linear program in Figure 3.6 constitute an IIS because the removal of any one of the two constraints makes the linear program feasible. There are several efficient algorithms available to detect IISs in a set of constraints. We used the *Elastic Filtering algorithm* described in [64]. The Elastic Filtering algorithm takes as input a set of linear constraints and identifies an IIS in these constraints (if one exists). An infeasible linear program may have more than one IISs in it, and the elastic filtering algorithm is guaranteed to find at least one of these IISs. To produce a feasible linear program from an infeasible linear program, we may be required to run the elastic filtering algorithm several times; each run identifies and removes an IIS and produces a smaller linear program that can further be examined for presence of IISs.

Figure 3.7 depicts our approach to obtaining a set of feasible linear constraints from a set of infeasible linear constraints. We first examine the input set C, to find out if it is feasible; if so, it does not contain IISs, and C can be used as the set of constraints in the linear-program formulation. If C

turns out to be infeasible, then it means that there is a subset of C that forms one or more IISs. This subset is denoted by C' in Figure 3.7. The elastic filtering algorithm, over several runs, identifies and removes the subset C' from the set of constraints. The resultant set $C - C'$ is feasible. We then set the values of the max and min variables appearing in C' to ∞ and $-\infty$, respectively. We do so because we cannot infer the values of these variables using linear programming, and hence setting these variables to infinite values is a conservative approach. The variables whose values are now infinite may appear in the set of constraints $C - C'$. The scenario is now similar to taint analysis, where we had some constraint variables whose values were infinite, and we had to identify and remove the constraint variables that were "tainted" by the infinite variables. Therefore, we run steps (2)-(13) of the taint-analysis algorithm (Figure 3.4) with Infset as the constraint variables that appear in C'. This step results in further removal of constraints, which are denoted by $C'' \subseteq C - C'$ in Figure 3.7. The set of constraints after removal of C'', denoted by D in Figure 3.7, satisfies the property that all max variables appearing in it have finite lower bounds, and all min variables have finite upper bounds. Moreover, D is feasible, and will only yield optimal solutions when solved as a linear program with the objective functions described earlier. Hence, we solve the linear program that consists of the set of constraints in D and a suitable objective function.

Figure 3.8 summarizes our approach to constraint resolution using linear programming. Steps (1)-(10) of the algorithm describe the transformation that removes the IISs, while step (11) performs the taint analysis to obtain a set of constraints that can be used in the linear program formulation.

3.2.2   IMPLEMENTATION

We implemented the above algorithm by extending the commercially available package SoPlex [257; 258]. SoPlex is a linear program solver; we extended it by adding IIS detection and taint analysis. In practice, linear program solvers work much faster when the constraints have been *presolved*. Presolving is a method by which constraints are simplified *before* being passed to the solver. In several cases, we can make simple inferences about the constraints; for instance, if $x \geq 5$ is the only constraint involving x, and we wish to minimize x, it is clear that x is 5. Several such techniques are described in the literature [22]; we incorporated some of them in our solver.

3.3   SOLVING CONSTRAINT SYSTEMS HIERARCHICALLY

In the previous section, we described an approach that used linear programming to determine bounds on the constraint variables. When the linear program was infeasible, we detected and removed IISs and solved a feasible subset of the constraints. In this section, we present an alternate approach for solving a set of constraints that handles infeasible sets of constraints in a different way. This approach was also developed independently by Rugina and Rinard in [222]. The idea behind this approach is to decompose the set of constraints into smaller subsets, and solve each subset separately. We do so by constructing a directed acyclic graph (DAG), each of whose members is a set of constraints, and solve each member in the order that it appears in a topological sort of the DAG.

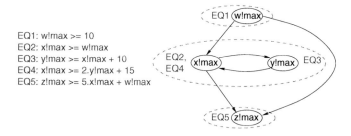

EQ1: w!max >= 10
EQ2: x!max >= w!max
EQ3: y!max >= x!max + 10
EQ4: x!max >= 2.y!max + 15
EQ5: z!max >= 5.x!max + w!max

Figure 3.9: Constraint Dependency Graph—an example.

To construct such a DAG, we first identify sets of constraints such that each member of the set depends on the other members of the set either directly or indirectly. Consider for instance, the constraints shown on the left in Figure 3.9. Constraint EQ3 gives a lower bound for the variable y!max based on the value of x!max. However, the value of x!max itself has a lower bound determined by EQ2 and EQ4. Thus EQ3 "depends" on EQ2 and EQ4.

To formalize the notion of dependency, we construct a graph whose vertices are the constraint variables in the set of equations. We associate the vertex corresponding to a variable x with all constraints in which x appears on the LHS. We draw an edge from a vertex y to a vertex x if there is a constraint that has y on the RHS and x on the LHS. We then identify *Strongly Connected Components (SCCs)* in this graph. The set of constraints associated with the vertices in an SCC are defined to be dependent upon each other. Figure 3.9 shows the constraints associated with each vertex; the SCCs are identified using dotted lines. EQ2, EQ3 and EQ4 are dependent on each other.

Recall that if we coalesce the SCCs in a graph, then the resulting graph is a DAG. The topological sort of the DAG naturally defines a hierarchy in the DAG. Hence, we consider each SCC in topologically sorted order and solve the constraints associated with that SCC. Each SCC consists of a set of linear constraints, and we formulate a linear program to minimize (maximize) each max (min) variable that appears in the set of constraints just as we did in Section 3.2.

If the set of constraints C in an SCC are found to be infeasible, we can immediately set all max and min variables appearing on the LHS of each constraint in C to ∞ and $-\infty$, respectively. This approach does not require us to identify and remove IISs in C. This is because an IIS-detection algorithm combined with the taint analysis that follows IIS detection, denoted by steps (1)-(11) of Figure 3.8, would remove all the constraints in C and set the variables appearing on the LHS in each constraint in C to infinite values. This can be attributed to the fact that each constraint in C is dependent on at least one more constraint in C; consequently, setting any LHS variable to an infinite value will result in the LHS variables of all constraints getting infinite values. Hence, the approach of solving a set of dependent constraints together obviates the need for IIS detection and elimination.

Once we have the values for the constraint variables that appear in an SCC, we can substitute these values in the constraints that are associated with the children of the SCC. Once all the SCCs have been solved, the values for all the constraint variables in the set of constraints becomes available.

A few points are worth noting with respect to this solver:

• Constraint simplification by substituting available values presents an opportunity to avoid calling an LP solver if the simplification makes the constraints amenable to presolving. For instance, for the set of constraints shown in Figure 3.9, we can infer that the value of `w!max` is 10 without having to invoke a linear program solver. This value can be substituted in `EQ2` and `EQ5`, thus simplifying these constraints. Similarly, we can infer the value of `z!max` once the value of `x!max` is available.

• The IIS-detection-based approach for handling infeasibility is an approximation algorithm. It may remove more constraints than are actually required to make the constraints feasible; as a result, more constraint variables than necessary may be set to ∞/-∞. It can be shown that the solution obtained by the hierarchical solver is precise, in the sense that it sets the fewest number of constraint variables to ∞/-∞. Furthermore, when the linear program is feasible, this solver produces the same solution as obtained by the algorithm in Figure 3.8. This gives rise to a trade-off, i.e., the user can choose between the hierarchical solver, which solves more (but smaller) linear programs, the solutions to which are mathematically precise, or choose the algorithm from Figure 3.8, which may be imprecise, but is more efficient. In our experiments, we noted that the approach from Section 3.2 can be up to 3 times faster than the hierarchical solver, while sacrificing the precision of only 5% of the constraint variables.

• Because we have broken down the problem into one of solving small sets of constraints, we could use a different solver for each set of constraints. Some kinds of constraint systems have fast solvers, for instance, the problem of finding a solution to a set of difference constraints can be formulated as a shortest-path problem [74].

• Lastly, for very large constraint systems, one could envision solving the SCCs at the same depth in parallel. Thus, a DAG with depth D can be solved in D steps.

3.4   ADDING CONTEXT SENSITIVITY

The constraint-generation process described in Section 3.1 was context-insensitive. When we generated the constraints for a function, we considered each call-site as an assignment of the actual-in variables to the formal-in variables, and the return from the function as an assignment of the formal-out variables to the actual-out variables. As a result, we merged information across call-sites, thus making the analysis imprecise. In this section, we describe how to incorporate context sensitivity.

Constraint inlining is similar in spirit to inlining function bodies at call-sites. Observe that in the context-insensitive approach, we lost precision because we treated *different* call-sites to a function identically, i.e, by assigning the actual-in variables at each call-site to the *same* formal parameter.

Constraint inlining alleviates this problem by creating a fresh instance of the constraints of the called function at each call-site. In other words, at each call-site to a function, we produce the constraints for the called function with the local variables and formal variables renamed uniquely for that call-site. This is illustrated in the example below, which shows some of the constraints for the function `copy_buffer` from Figure 3.2 specialized for the call-site at line (7):

```
copy!alloc!max₁ ≥ buffer!used!max₁ - 1
copy!used!max₁ ≥ buffer!used!max₁
copy!used!min₁ ≤ buffer!used!min₁
copy_buffer$return!used!max₁ ≥ copy!used!max₁
copy_buffer$return!used!min₁ ≤ copy!used!min₁
```

Context-sensitivity can be obtained by modeling each call-site to the function as a set of assignments to the renamed instances of the formal variables. The actual-in variables are assigned to the *renamed* formal-in variables, and the *renamed* formal-out variables are assigned to the actual-out variables. As a result, there is exactly one assignment to each renamed formal-in parameter of the function, which alleviates the problem of merging information across different calls to the same function.

Some of the constraints for the call-site to `copy_buffer` at line (7) in Figure 3.2 are shown below:

```
buffer!used!max₁ ≥ header!used!max
buffer!used!min₁ ≤ header!used!min
cc1!used!max ≥ copy_buffer$return!used!max₁
cc1!used!min ≤ copy_buffer$return!used!min₁
```

With this approach to constraint generation, we obtain the values 2047 and 2048 for `cc1!alloc!max` and `cc1!used!max`, respectively, while `cc2!alloc!max` and `cc2!used!max` get the values 1023 and 1024, respectively, which is an improvement over the values reported in Figure 3.5.

We have implemented this approach because it requires only minimal changes to the constraint generation process that we have already described. However, it also has some shortcomings:

• It does not handle recursive function calls: inlining cannot work in the presence of recursion.

• The number of constraint variables in the constraints with context sensitivity may be exponentially larger than the number of constraints in their context-insensitive counterpart. As a result, we do not expect this technique to scale well to large programs.

These drawbacks can be overcome through the use of *summary constraints*. Summary constraints summarize the effect of a function call in terms of the constraint variables representing global variables and formal parameters of the called function. Once the summary constraints of a function are available, we can obtain context sensitivity by substituting actual parameters in place of the formal parameters in the summary constraints. This approach is described next.

3.4.1 SUMMARY CONSTRAINTS

The approach described in this section addresses the shortcomings of constraint inlining – namely, the method described here handles recursion and does not result in a large number of variables. The basic idea is to summarize the effect of a function call using a set of constraints expressed only in terms of constraint variables that denote the global program variables and formal parameters of the called function. We refer to such constraints as *summary* constraints. For instance, consider the function copy_buffer shown in Figure 3.2. Figure 3.10 shows a subset of constraints (only those involving max variables) generated for copy_buffer, and it shows the corresponding summary constraints. Notice that the summary constraints are produced in terms of constraint variables that denote the formal parameters of copy_buffer. The program statements responsible for generating each constraint are also shown.

Subset of Constraints Generated by copy_buffer
(1) copy!alloc!max ≥ buffer!used!max − 1 (*from line* 17)
(2) copy!used!max ≥ buffer!used!max (*from line* 18)
(3) copy_buffer$return!alloc!max ≥ copy!alloc!max (*from line* 19)
(4) copy_buffer$return!used!max ≥ copy!used!max (*from line* 19)
Equivalent Summary Constraints
(A) copy_buffer$return!alloc!max ≥ buffer!used!max − 1
(B) copy_buffer$return!used!max ≥ buffer!used!max

Figure 3.10: Summary constraints for copy_buffer.

To summarize the effect of a function call, we must eliminate all constraint variables that correspond to local variables of the called function. This results in a new set of constraints for the called function in terms of constraint variables that correspond to formal parameters and global program variables. There are several variable elimination techniques available for linear constraint systems, the most common one being the *Fourier–Motzkin elimination* method. The Fourier-Motzkin method takes as input a set of constraints C and a set of variables V that must be retained in the summary constraints. It then iteratively eliminates the variables not in V. For example, for the constraints shown in Figure 3.10 the Fourier-Motzkin method would eliminate copy!alloc!max by combining constraints (1) and (3) to produce constraint (A). Similarly, it would eliminate copy!used!max by combining constraints (2) and (4) to produce constraint (B).

The Fourier-Motzkin variable-elimination method works with affine constraints, and in the worst case, can result in the generation of a large number of constraints. Specifically, if we want to eliminate a variable v from a set of constraints, where m constraints use v and n constraints define v, the output could have as many as $m \cdot n$ constraints. This problem can be partially alleviated by eliminating constraints that are implied by other constraints.

Our observation, however, is that most of the constraints that are generated by our tool are difference constraints, i.e., two variable constraints of the form $v_1 \geq v_2 \pm w$ (where v_1 and v_2 are either both max variables or both min variables). For instance, about 98.8% of the constraints generated by our tool for sendmail-8.7.6 were difference constraints. Variable elimination for difference constraints reduces to an all-pairs shortest-path or longest-path problem on a graph formed by the constraints. Hence, we will restrict our exposition to the case when a function generates only difference constraints. When we consider a function that only generates difference constraints, the constraint subsystem involving the max variables is completely disjoint from the constraint subsystem involving the min variables. This means that we can produce summary constraints for each of these subsystems independently.

First, consider a function that does not call other functions or only calls those functions for which summary functions are available. The function `copy_buffer` from Figure 3.2 is an example of such a function because it only makes calls to `strcpy` and `malloc`, and we have summary functions for both of these.

To produce summary constraints for a set of constraints C of such a function in terms of a set of variables V (the set of constraint variables for formal-parameters and globals of the function), we construct a graph to denote the constraints in C. The vertices of this graph are the constraint variables that appear in C. For a constraint of the form $v_1 \geq v_2 + w$, where v_1 and v_2 are `max` variables, we draw an edge with weight w from v_2 to v_1. Because there may be several constraints that relate v_1 and v_2, the edge is assigned a weight equal to the greatest difference between these variables. For each constraint of the form $v_1 \geq w$, we draw an edge with weight w from a dummy "zero" variable v_0 to v_1. For instance, the graph of the constraints involving the `max` variables for the function `copy_buffer` is shown in Figure 3.11 (the variable `buffer!alloc!max` was not involved in any constraints generated by `copy_buffer` and hence is not shown in the graph). The problem of generating summary functions now reduces to finding the *longest* path between each pair of vertices in V. Intuitively, the longest path length is the maximum difference between the two variables. Hence, if the longest path length from v_1 to v_2 is a, we would generate the constraint $v_2 \geq v_1 +$ a. The all-pairs shortest-path problem for vertices in V can be solved using well-known techniques (such as the Floyd-Warshall algorithm [74]). Thus, for `copy_buffer`, the graph shown in Figure 3.11 yields the constraints shown in Figure 3.10.

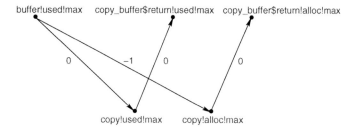

Figure 3.11: Graph for Summary Constraint Production.

An analogous construction for the `min` variables helps produce the summary constraints for the constraints consisting of the `min` variables. In this case, a constraint $v_1 \geq v_2 + w$ would result in an edge with weight w from v_2 to v_1, where v_1 and v_2 are `min` variables. However, in this case we would be required to find the *shortest* path between each pair of vertices in V.

We can now use the summary constraints computed for `copy_buffer` in `main` to make the calls to `copy_buffer` context-sensitive. This is depicted in Figure 3.12. This figure shows the portion of the constraint graph of `main` from Figure 3.2 pertaining to the constraints generated at line (10). The dotted edge originating from `buf!used!max` denotes the assignment of `buf!used!max` to `buffer!used!max`, while the dotted edges incident on `cc2!used!max` and `cc2!alloc!max` denote the assignment statements from the formal-out constraint variables of `copy_buffer` to the actual-out constraint variables. The dotted edges from `buffer!used!max` to `copy_buffer$return!alloc!max` and `copy_buffer$return!used!max` denote the summary constraints (A) and (B) from Figure 3.10, respectively, and are computed by obtaining the pairwise longest paths from Figure 3.11. The bold edges denote the substitution of the actual variables in place of the formal parameters in the summary constraints. When we generate constraints for `main`, we only generate the constraints pertaining to the bold lines shown in Figure 3.12. Hence, the call to `copy_buffer` at line (10) in the program in Figure 3.2 would result in the constraints

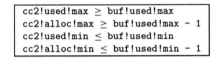

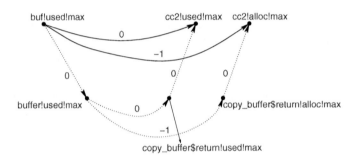

Figure 3.12: Obtaining Context-Sensitivity.

The above technique can be summarized as follows:
• Inspect the call-graph of the program, identify SCCs in it, and coalesce all the nodes belonging to an SCC.
• The resultant graph is a DAG; compute summary constraints in reverse topological order of the DAG. For each function that calls other functions, summarize the effect of the call by substituting the actual variables in place of the formal parameters of the called function.

3.5   EXPERIENCE WITH THE TOOL

We tested our prototype implementation on several popular commercially used programs. In each case, the tool produced several warnings; we used these warnings, combined with Codesurfer features such as slicing, to check for real overruns. We tested to see if the tool discovered known overruns documented in public databases such as `bugtraq` [3] and CERT [4], and we also checked to see if any overruns that were previously unreported were discovered. We report our experience with `wu-ftpd` and `sendmail`.

All our experiments were performed on a machine with a 3GHz P4 Xeon processor machine with 4GB RAM running Debian GNU/Linux 3.0. We used Codesurfer version 1.8 for our experiments, the gcc-3.2.1 compiler for building the programs, and glibc version 2.2.4 for macro-expansion. Codesurfer implements several pointer-analysis algorithms; in each case, we performed the experiments with a field-sensitive version of Andersen's analysis [21] that uses the common-initial-prefix technique of Yong and Horwitz [260] to deal with structure casts. We configured the tool to use the hierarchical solver described in Section 3.3 for constraint resolution (so the values obtained will be precise) and produce constraints in a context-insensitive fashion.

3.5.1 WU-FTP DAEMON

We tested two versions of the wu-ftp daemon, a popular file transfer server. Version 2.5.0 is an older version with several known vulnerabilities (see CERT advisories CA-1999-13, CA-2001-07 and CA-2001-33), while version 2.6.2 is the current version with several security patches that address the known vulnerabilities.

3.5.1.1 wu-ftpd-2.6.2

wu-ftpd-2.6.2 has about 18K lines of code and produced 178 warnings when examined by our tool. Upon examining the warnings, we found 14 previously unreported overruns; we will describe one of these in detail.

The tool reported a potential overrun on a buffer pointed to by accesspath in the procedure read_servers_line in rdservers.c, where as many as 8192 bytes could be copied into the buffer for which up to 4095 bytes were allocated. Figure 3.13 shows the code snippet from read_servers_line that is responsible for the overrun.

```
int read_servers_line (FILE *svrfp,
                       char *hostaddress,
                       char *accesspath){
    static char buffer[BUFSIZ];
    ...
    while (fgets(buffer, BUFSIZ, svrfp)){
      ...
      if ((hp = gethostbyname(hcp))){
        struct in_addr in;
        memmove(&in, hp->h_addr, sizeof(in));
        strcpy(hostaddress, inet_ntoa(in));
      }
      else
        strcpy(hostaddress, hcp);

      strcpy(accesspath, acp);
    }
}
```

Figure 3.13: Code snippet from wu-ftpd-2.6.2.

The fgets statement may copy as many as 8192 (BUFSIZ) bytes into buffer, which is processed further in this function. As a result of this processing, acp and hcp point to locations inside buffer.

By an appropriate choice of the contents of `buffer`, one could make `acp` or `hcp` point to a string buffer as long as 8190 bytes, which could result in an overflow on the buffer pointed to either by `accesspath` or `hostname`, respectively.

The procedure `read_servers_line` is called at several places in the code. For instance, it is called in the main procedure in `ftprestart.c`, where `read_servers_line` is called with two local buffers, `hostaddress` and `configdir`, which have been allocated 32 bytes and 4095 bytes, respectively. This call reads the contents of the file `_PATH_FTPSERVERS`, which typically has privileged access. However, in non-standard and unusual configurations of the system, `_PATH_FTPSERVERS` could be written to by a local user. As a result, the buffers `hostaddress` and `configdir` can overflow based on a carefully chosen input string, possibly leading to a local exploit. The use of a `strncpy` or `strlcpy` statement instead of the unsafe `strcpy` in `read_servers_line` rectifies the problem.

Some other new overruns that were detected by the tool were:
- An unchecked `sprintf` in `main` in the file `ftprestart.c` could result in 16383 bytes being written into a local buffer that was allocated 4095 bytes.
- Another unchecked `sprintf` in `main` in the file `ftprestart.c` could result in 8447 bytes being written into a local buffer that was allocated 4095 bytes.
- An unchecked `strcpy` in `main` in the file `ftprestart.c` could result in 8192 bytes being written into a local buffer that was allocated 4095 bytes.

In each of the above cases, a carefully chosen string in the file `_PATH_FTPACCESS` can be used to cause the overrun. As before, `_PATH_FTPACCESS` typically has privileged access but could be written to by a local user in non-standard configurations. We contacted the `wu-ftpd` developers [154], and they have acknowledged the presence of these bugs in their code, and are in the process of fixing the bugs (at the time of writing this paper).

3.5.1.2   wu-ftpd-2.5.0

`wu-ftpd-2.5.0` has about 16K lines of code; when analyzed by our tool, it produced 139 warnings. Figure 3.14 shows a screenshot of the GUI that the tool provides for the user to surf the warnings. Each of the warnings shown is a "hot" link and is linked back to the line of source code that is responsible for the warning. Consider the first warning shown in the figure; it depicts that the tool found a potential overrun on buffer `buf` in the procedure `vreply` in the file `ftpd.c`. It also shows two possible locations where the overrun could have occurred – an `snprintf`, and an `sprintf` statement.

The first location where a potential overrun was found, a call on `snprintf`, contains:

```
snprintf(buf + (n ? 4 : 0),
        n ? (sizeof(buf)-4) : sizeof(buf),
        "%s", fmt);
```

Clearly, no more than `sizeof(buf)` bytes are written into `buf`, and hence this statement is safe. However, because the tool ignores control flow, this statement is modeled as though `sizeof(buf)` bytes could be written at the location `buf + 4`, which causes the tool to report that as many as 8196 bytes could be written into `buf` for which 8192 bytes where allocated. As a result, this warning is

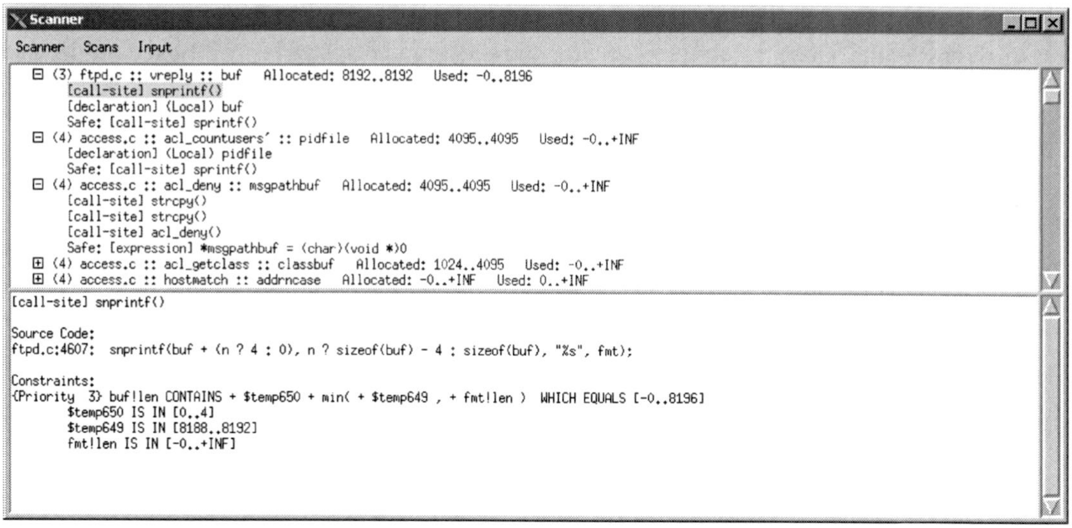

Figure 3.14: A screenshot from the `wu-ftpd-2.5.0` analysis.

a false positive. The second location associated with this warning, a call on `sprintf` , turns out to be safe because it copies only 16 bytes into the 8192 byte array `buf`. The tool inferred this from the constraints, and hence this statement was marked "safe" as shown in the Figure 3.14.

We analyzed the warnings to check for a widely exploited overrun reported in CERT advisory CA-1999-13. The buffer overrun is on a globally declared buffer `mapped_path` in the procedure `do_elem` in the file `ftpd.c`. It was noted in [156] that the overrun was due to a statement `strcat(mapped_path, dir)`, where the variable `dir` could be derived (indirectly) from user input. As a result, it is possible to overflow `mapped_path` for which 4095 bytes were allocated. Our tool reported the range for `mapped_path!used` as $[0..+\infty]$, while `mapped_path!alloc` was $[4095..4095]$. We note that `strcat(dst, src)` would be modeled as four linear constraints by our tool:

$$
\begin{aligned}
dst'!used!max &\geq dst!used!max + src!used!max \\
dst!used!max &\geq dst'!used!max \\
dst'!used!min &\leq dst!used!min + src!used!min \\
dst!used!min &\leq dst'!used!min
\end{aligned}
$$

The first two constraints make the linear program infeasible, as explained in Section 3.2 and result in `dst!used!max` being set to $+\infty$. Hence, in `wu-ftpd-2.5.0`, `mapped_path!used!max` will be set to $+\infty$, and the tool would have reported the same range even in the absence of an overrun. We used Codesurfer's program slicing feature to confirm that `dir` could be derived from user input. We found that the procedure `do_elem`, one of whose parameters is `dir`, is called from the procedure `mapping_chdir`. This function is in turn called from the procedure `cmd`, whose input arguments can be controlled by the user. This shows the importance of providing the end user with several program-analysis features. These features, such as program slicing and display of control and data-dependence predecessors, which are part of Codesurfer, aid the user of the tool in understanding the source code better and hence in locating the source of the vulnerability.

3.5.2   SENDMAIL

Sendmail is a very popular mail-transfer program. We analyzed sendmail-8.7.6, an old version that was released after a thorough code audit of version 8.7.5. However, this version has several known vulnerabilities. We also analyzed sendmail-8.11.6; in March 2003, two new buffer-overrun vulnerabilities were reported in the then current sendmail version. We note that sendmail-8.7.6 and sendmail-8.11.6 are vulnerable to these overruns as well.

3.5.2.1   sendmail-8.7.6

sendmail-8.7.6 has about 38K lines of code; when analyzed by our tool, it produced 295 warnings. Due to the large number of warnings, we focused on scanning the warnings to detect some known overruns.

Wagner *et al.* use BOON [253] to report an off-by-one bug in sendmail-8.9.3 where as many as 21 bytes, returned by a function queuename, could be written into a 20-byte array dfname. Our tool identified four possible program points in sendmail-8.7.6 where the return value from queuename is copied using strcpy statements into buffers that are only 20 bytes long. As in [253], our tool reported that the return value from queuename could be up to 257 bytes long, and further manual analysis was required to determine that this was in fact an off-by-one bug. Another off-by-one bug was reported by our tool where the programmer mistakenly allocated only 3 bytes for the buffer delimbuf which stored "\n\t ", which is 4 bytes long including the end-of-string character.

3.5.2.2   sendmail-8.11.6

sendmail-8.11.6 is significantly larger than version 8.7.6 with 68K lines of code; when we ran our tool, it produced 453 warnings. We examined the warnings to check if the tool discovered the new vulnerabilities reported in March 2003.

One of these vulnerabilities is on a function crackaddr in the file headers.c, which parses an incoming e-mail address string. This function stores the address string in a local static buffer called buf that is declared to be MAXNAME + 1 bytes long, and performs several boundary-condition checks to see that buf does not overflow. However, the condition that handles the angle brackets (<>) in the From address string is incorrect, thus leading to the overflow [204].

Our tool reported that bp, a pointer to the buffer buf in the function had bp!alloc!max = +∞ and bp!used!max = +∞, thus raising a warning. However, the reason bp!alloc!max and bp!used!max were set to +∞ was because of several statements that perform pointer arithmetic in the body of the function. In particular, the statement bp-- results in bp!alloc!max = +∞ and bp!used!max = +∞. Hence, this warning would have existed even if the boundary-condition checks were correct.

3.5.2.3 Talk Daemon

The talk daemon program, a popular UNIX communication facility, derived from the current FreeBSD release is about 900 lines of code, and produced just 4 warnings by our tool. Upon furthur investigation, we found that all the 4 warnings were false alarms; however, one of the warnings was particularly interesting.

The tool reported that as many as 33 bytes could be copied into a buffer pointed to by `tty` which was allocated 16 bytes. The source code responsible for this warning is shown in Figure 3.15.

On our platform, `UT_LINESIZE` macro-expanded to 32, as a result of which the tool reported the overrun. However, we discovered that when we used the FreeBSD header files for macro-expansion, `UT_LINESIZE` was 8, and hence the warning was suppressed.

This example serves to demonstrate the use of our tool to determine whether an application is vulnerable on a particular platform. For instance, the talk daemon program would have been vulnerable to the aforementioned buffer overrun on our platform.

```
struct utmp
  char   ut_line[UT_LINESIZE];
  ...
};

int find_user(const char *name, char *tty)
  struct utmp ubuf;
  char line[sizeof(ubuf.ut_line) + 1];

  while (fread((char *) &ubuf, sizeof ubuf ..))
    strncpy(line, ubuf.ut_line,
                    sizeof(ubuf.ut_line));
    line[sizeof(ubuf.ut_line)] = '\0';
  ...
      if (...)
      ...
      (void) strcpy(tty, line);
      ...
```

Figure 3.15: Code Snippet from Talk Daemon.

3.5.2.4 Telnet Daemon from `kth-kerberos-4.0.0`

We tested the Telnet daemon program from the KTH release of `kerberos-4.0.0` (circa 1995). Telnet daemon has about 9400 lines of code and produced 40 warnings when analyzed by the tool. The tool identified an interesting bug: it reported that as many as 256 bytes could be copied into `terminaltype`, which points to a buffer only 41 bytes long. We found that the bug was due to a `strncpy` statement in `getterminaltype` in the file `telnetd.c`:

`strncpy(terminaltype, first, sizeof(first))`

Note that `strncpy` was meant to be a "safe" function but was used in an unsafe way – the programmer mistakenly set the number of bytes to be copied into the destination buffer equal to the size of the source buffer, thus rendering the `strncpy` statement equivalent to its "unsafe" counterpart `strcpy`.

	`wu-ftpd-2.6.2`	`sendmail-8.7.6`
CODESURFER	12.54 sec	30.09 sec
GENERATOR	74.88 sec	266.39 sec
TAINT	9.32 sec	28.66 sec
LPSOLVE	3.81 sec	13.10 sec
HIERSOLVE	10.08 sec	25.82 sec
Number of Constraints Generated		
PRE-TAINT	22008	104162
POST-TAINT	14972	24343

Figure 3.16: Performance of the tool.

3.5.3   PERFORMANCE

Figure 3.16 contains representative numbers from our experiments with `wu-ftpd-2.6.2` and `sendmail-8.7.6`. All timings are wall-clock times and are an average over 4 runs; CODESURFER denotes the time taken by Codesurfer to analyze the program, GENERATOR denotes the time taken for constraint generation, while TAINT denotes the time taken for taint analysis. The constraints produced can be resolved in one of two ways; the rows LPSOLVE and HIERSOLVE report the time taken by the solvers from Section 3.2 and Section 3.3, respectively. The number of constraints output by the constraint generator is reported in the row PRE-TAINT, while POST-TAINT denotes the number of constraints after taint analysis.

These results serve to demonstrate the trade-off between performance and precision of the hierarchical solver versus the IIS-detection-based solver from Section 3.2. While the IIS-detection-based approach is much faster, it is not mathematically precise. However, we found that it is a good approximation to the solution obtained by the hierarchical solver. In case of `wu-ftpd-2.6.2`, fewer than 5% of the constraint variables and, in the case of `sendmail-8.7.6`, fewer than 2.25% of the constraint variables obtained imprecise values when we used the IIS-detection-based approach.

3.5.4   ADDING CONTEXT SENSITIVITY

We report here our experience with using context-sensitive analysis on `wuftpd-2.6.2` using both the constraint inlining approach and the summary-constraints approach. To measure the effectiveness of each approach, we will count the number of range variables that were refined in comparison to the corresponding ranges obtained in a context-insensitive analysis. Recall that the value of a range variable `var` is given by the corresponding constraint variables `var!min` and `var!max` as [`var!min..var!max`]. We chose this metric because, as explained in Section 3.1.5, the tool uses the values of the ranges to produce diagnostic information, and more precise ranges will more precise diagnostic information.

The context-insensitive analysis on `wuftpd-2.6.2` yields values for 7310 range variables. Using the summary-constraints approach, we found that 72 of these range variables obtained more precise values. Note that in this approach the number of constraint variables (and hence the number of range variables) is the same as in the context-insensitive analysis. However, the number of constraints may change, and we observed a 1% increase in the number of constraints. This change can be attributed to the fact that summarization introduces some constraints (the summaries), and removes some constraints (the old call-site assignment constraints).

The constraint-inlining approach, on the other hand, leads to a $5.8\times$ increase in the number of constraints and a $8.7\times$ increase in the number of constraint variables (and hence the number of range variables). This can be attributed to the fact that the inlining approach specializes the set of constraints at each callsite. In particular, we observed that the 7310 range variables from the context-insensitive analysis were specialized to 63704 range variables based on calling context. We can count the number of range variables for which more precise values were obtained in two ways:

• Out of the 63704 specialized range variables, 7497 range variables had obtained more precise values than the corresponding unspecialized range variables.

• Out of the 7310 unspecialized range variables, 406 range variables had obtained more precise values in at least one calling context.

As noted earlier, the constraint-inlining approach returns more precise information than the summary-constraints approach. To take a concrete example, we consider the program variable msgcode (an integer), which is the formal parameter of a function pr_mesg in the file access.c in wu-ftpd-2.6.2. The function pr_mesg is called from several places in the code with different values for the parameter msgcode. The summary-constraints approach results in the value [530..550] for the range variable corresponding to msgcode. However, the constraint-inlining approach refines these ranges – for instance, it is able to infer that pr_mesg is always called with the value 530 from the function pass in the file ftpd.c.

3.5.5   EFFECTS OF POINTER ANALYSIS

As observed in Section 3.1, we were able to reduce false negatives through the use of pointer analysis. The tool is capable of handling arbitrary levels of dereferencing. For instance, if p points to a pointer to a structure s, the pointer-analysis algorithms correctly infer this fact. Similarly, if p and q are of type char** (i.e., they point to pointers-to-buffers), the constraints for a statement such as strcpy(*p, *q) would be correctly modeled in terms of the points-to sets of p and q (recall that we generated constraints in terms of pointers to buffers rather than buffers themselves).

To observe the benefits of pointer analysis, we generated constraints with the pointer analysis algorithms turned off. Because fewer constraints will be generated, we can expect to see fewer warnings; in the absence of these warnings, false negatives may result. We observed a concrete case of this in sendmail-8.7.6. When we generated constraints without including the results of pointer analysis, the tool output only 251 warnings (as opposed to 295 warnings). However, this method suppressed the warning on the array dfname, so the tool missed the off-by-one bug that we described earlier. A closer look at the code in procedure queuename revealed that in the absence of points-to facts, the tool failed to generate constraints for the statement

```
snprintf(buf, sizeof buf, "%cf%s",
                              type, e- >e_id)
```

in the body of `queuename` because points-to facts for the variable e, which is a pointer to a structure, were not generated.

We note that BOON [253] identified this off-by-one bug because of a simple assumption made to model the effect of pointers, i.e., BOON assumes that any pointer to a structure of type T can point to all structures of type T. While this technique can be effective at discovering bugs, the lack of precise points-to information will lead to a larger number of false alarms.

3.5.6   SHORTCOMINGS

While we found the prototype implementation a useful tool to audit several real-world applications, we also noted several shortcomings.

First, the flow insensitivity of the analysis meant that it reports false positives. Through the use of slicing, we were able to identify the false positives, nevertheless it was a manual and often painstaking procedure. By transitioning to a Static Single Assignment (SSA) representation [84] of the program, we can add a limited form of flow sensitivity to the program. This will result in a large number of constraint variables. Fortunately, we have observed that the solvers readily scale to large linear programs with several thousand variables.

Second, by modeling constraints in terms of pointers to buffers rather than buffers, we can miss overruns, thus leading to false negatives [253]. However, the reason we did so was because the pointer-analysis algorithms themselves were flow- and context-insensitive, and generating constraints in terms of buffers would have resulted in a large number of constraints and, consequently, a large number of false positives. By transitioning to "better" pointer analysis algorithms, we can model constraints in terms of buffers themselves, thus eliminating the false negatives.

3.6   RELATED WORK

Several static techniques have been proposed to mitigate the problem of buffer overruns. Wagner *et al.* [252; 253] proposed a tool, BOON, similar in spirit to ours to detect buffer overruns in C source code. However, unlike our tool, BOON does not employ pointer analysis and does not provide a way to enhance the context-sensitivity of the analysis. Larochelle and Evans [156] propose an extension to LCLint that uses semantic information from annotations in the program to make inferences on buffer bounds. The tool works like a compiler and produces warnings by making inferences based on the annotations. Xi and Pfenning [259] propose an extension to ML that supports type annotations. These annotations are then used to determine the type safety of programs. However, in both these techniques, the onus is on the user to provide correct annotations. As a result, analyzing large legacy applications without annotations becomes almost impossible. Dor *et al.* [98] propose a tool (CSSV) that aims to find all buffer overflows with just a few false positives. The basic idea is to convert the C program into an inrgeteger program and use a conservative static-analysis algorithm that can check the correctness of integer manipulations. The analysis is performed on a per-procedure basis; to extend the analysis interprocedurally, they use the concept of *contracts*, which are a set of pre-conditions and post-conditions of a procedure, along with side-effect information. The number of false positives generated depends on the accuracy of the contracts, which are typically provided by the user. They also discuss techniques whereby conservative, user-supplied contracts can be automatically refined. Rugina and Rinard [222] propose a technique based on linear programming that infers symbolic upper and lower bounds on arrays. They deal with infeasible linear programs by using a solver similar to the hierarchical-solver approach presented in Section 3.3. They use a flow- and context-sensitive program analysis to detect several programming errors such as array out-of-bounds errors and race conditions. However, the techniques in [98; 222] have not been tested on large programs, and may scale poorly. For instance, CSSV took > 200 seconds to analyze a string-manipulation program with a total of about 400 lines of code.

There are a suite of dynamic techniques that help prevent stack-smashing attacks. Stack-guard [80] detects changes to the return address by placing a "canary" word on the stack. RAD [65] defends the return address by storing it in a repository and checking the return address against the repository before the function returns. Both these techniques enhance the compiler to insert function prologues and epilogues that perform the checking. Prasad and Chiueh [211] propose a binary-rewriting technique that enhances binaries by incorporating the RAD mechanism; however, their technique suffers from imprecision while disassembling the binary. While these methods help in *detecting* and *preventing* attacks based on buffer overruns, they fail to *eliminate* the buffer overflows from the source code which is the goal around which our tool is built.

Static techniques have also been used to reduce the overhead of run-time checks. CCured [73; 200] is a program-transformation system that adds memory-safety guarantees to C programs by statically analyzing the source code and classifying pointers as safe or unsafe. Appropriate run-time checks are then inserted depending on the kind of the pointer (lightweight checks for safe pointers). CCured has been applied to several commercial applications with reasonable run-time overhead [73]. However, in some cases, such as systems software, the overhead of CCured could be as high as 87%. Bodik *et al.* present ABCD [45], which provides a way to eliminate frequently executed but redundant array-bounds checks for Java programs. This technique assumes the presence of run-time checks in the code, and provides a way to improve performance by removing some of the checks.

CHAPTER 4

Analyzing Security Policies

The main issues in access control are authentication, authorization, and enforcement. Identification of principals is handled by authentication, and authorization addresses the question "Should a request *r* by a specific principal *A* be allowed?" Enforcement addresses the problem of implementing the authorization during an execution. In this chapter, we first discuss classic topics related to access control. This leads us to a discussion of safety analysis for *Role Based Access Control (RBAC)*, a widely used model for access control. Later sections discuss analysis of policies in two trust management systems, SPKI/SDSI and RT.

Access control is a fundamental and pervasive security mechanism in use today. It controls which principals (users, processes, machines, etc.) have access to which resources in a system; for example, which files they can read, which programs they can execute, and how they share data with other principals. The specification and management of access-control policies is a challenging problem. A large number of security problems are caused by misconfigurations. Formal analysis and verification techniques can be used in access-control-policy specification and management. In formal analysis, one develops a specification of the policy objectives and a description of the access-control system, and verifies whether the policy objectives are achieved.

Managing access control is relatively easy when the authorization state is fixed. However, in any real system, there is always a need to change the authorization state; for example, users and objects are added and removed, users start sharing resources at one moment and stop such sharing later, and users' job functionalities change. It is this dynamic aspect that makes access control challenging. One fundamental problem dealing with the dynamic aspect of access control is safety analysis, first formulated by Harrison, Ruzzo, and Ullman [125] for the access-matrix model [153]. Safety analysis decides whether rights can be leaked to unauthorized principals in future states. Safety analysis was shown to be undecidable in the HRU scheme. Since then, safety analysis has been considered to be a fundamental problem in access control, and there has been considerable work on safety analysis in various contexts related to security [19; 20; 54; 144; 162; 163; 166; 189; 196; 226; 227; 228; 229; 243; 245; 246]. In recent years, the notion of *security analysis* was introduced. It generalizes safety analysis and asks the fundamental question whether an access-control system preserves a security policy invariant (which encodes desired security properties) across state changes.

We now define an abstract version of security analysis for general access-control schemes.

Definition 4.1 (*Access-Control Schemes*) An access-control scheme is modelled as a state-transition system $\langle \Gamma, Q, \vdash, \Psi \rangle$, in which Γ is a set of states, Q is a set of queries, Ψ is a set of state-change rules, and $\vdash: \Gamma \times Q \to \{true, false\}$ is called the entailment relation, determining whether a *query* is true or not in a given state. A *state*, $\gamma \in \Gamma$, contains all the information necessary for making access-control decisions at a given time. When a query, $q \in Q$, arises from an access request, $\gamma \vdash q$ means that the access corresponding to the request q is granted in the state γ, and $\gamma \not\vdash q$ means that the access corresponding to q is not granted. One may also ask queries other than those corresponding to a specific request, e.g., whether every principal that has access to a resource is an employee of the organization. Such queries are useful for understanding the properties of a complex access-control system.

A state-change rule, $\psi \in \Psi$, determines how the access-control system changes state. Given two states γ and γ_1 and a state-change rule ψ, we write $\gamma \mapsto_\psi \gamma_1$ if the change from γ to γ_1 is allowed by ψ, and $\gamma \stackrel{*}{\mapsto}_\psi \gamma_1$ if a sequence of zero or more allowed state changes leads from γ to γ_1. If $\gamma \stackrel{*}{\mapsto}_\psi \gamma_1$, we say that γ_1 is *ψ-reachable* from γ, or simply γ_1 is *reachable*, when γ and ψ are clear from the context.

Definition 4.2   (*Security Analysis in an Abstract Setting*) Given an access-control scheme $\langle \Gamma, Q, \vdash, \Psi \rangle$, a security-analysis instance takes the form $\langle \gamma, q, \psi \rangle$, where $\gamma \in \Gamma$ is a state, $q \in Q$ is a query, $\psi \in \Psi$ is a state-change rule. An instance $\langle \gamma, q, \psi \rangle$ asks whether for every γ_1 such that $\gamma \stackrel{*}{\mapsto}_\psi \gamma_1$, $\gamma_1 \vdash q$.

4.1   ACCESS-MATRIX-BASED SYSTEMS

Safety analysis was first formalized in the landmark paper by Harrison, Ruzzo, and Ullman [125]. We call the access-control scheme in [125] the HRU scheme.

THE HRU SCHEME

In a system based on the HRU scheme, each state $\gamma \in \Gamma$ is associated with a tuple $\langle S_\gamma, O_\gamma, M_\gamma[\ ] \rangle$, where S_γ and O_γ are finite sets of subjects and objects, respectively, that exist in γ, and $M_\gamma[\ ]$: $S_\gamma \times O_\gamma \to 2^{R_\psi}$ is an access matrix that maps a $\langle subject, object \rangle$ pair to a set of rights. R_ψ is the finite set of all such rights in the system, and we use 2^{R_ψ} to denote the powerset of R_ψ. We prefer to associate the finite set of rights, R_ψ, with the state-change rule ψ rather than the state γ because the set is fixed across all states for a system. We postulate that $S_\gamma \subset S$ and $O_\gamma \subset O$, where S and O are countable sets of all possible subjects and objects, respectively, that can exist in a state. Every subject is also an object, i.e., $S \subseteq O$ and $S_\gamma \subseteq O_\gamma$ in every state γ.

A state change in an HRU system is the successful execution of a command. The state-change rule ψ is specified by a finite set of rights, R_ψ, and a finite set of commands, where each command is of the following form.

$commandName(x_1, \ldots, x_a)$
$\qquad$ if $r_1 \in M[x_{i_1}, x_{j_1}] \wedge \cdots \wedge r_b \in M[x_{i_b}, x_{j_b}]$ then

primitive-operation$_1$

$\vdots$

primitive-operation$_c$

In the above command, a and c are positive integers, and b is a non-negative integer. The string *commandName* is the name of the command, and $x_1, \ldots, x_a$ is a list of parameters. The "if …then" portion checks for the presence of rights in cells in the (current instance of the) access matrix, and each check is of the form $r_k \in M[x_{i_k}, x_{j_k}]$ and is called a condition. When $b = 0$, we have a command with no conditions; when $b \leq 1$, we have a *mono-conditional* command; and when $b \leq 2$, we have a *bi-conditional* command. When $c = 1$, we have a *mono-operational* command. We require that the parameters $x_{i_1}, \ldots, x_{i_b}$ be instantiated with subjects (otherwise, an attempt at executing the command fails). We also require that $\{r_1, \ldots, r_b\} \subseteq R_\psi$ (the r_i's do not have to be distinct from one another). If we assume that γ is the state in which we attempt to execute the command, then the "if …then" check succeeds if and only if the rights are present in the corresponding cells of the access matrix in the state γ.

In the execution of a command, the "if …then" conditions are first evaluated. If any of them is not met, the command fails to execute. Otherwise, the primitive operations are executed in sequence. If the execution of a primitive operation succeeds, then the access matrix is altered in a corresponding way, as we discuss below for each kind of primitive operation. If the execution of a primitive operation fails, the access matrix is unaltered. Each primitive operation is of one of the following forms. In each of the following, we use $\langle S_\gamma, O_\gamma, M_\gamma[\,] \rangle$ to denote the state immediately before the execution of the primitive operation, and $\langle S_{\gamma'}, O_{\gamma'}, M_{\gamma'}[\,] \rangle$ to denote the state after immediately after the successful execution.

enter r into M[x, y] : when x is instantiated with s and y with o, this operation succeeds if and only if $s \in S_\gamma$ and $o \in O_\gamma$. The result is $M_{\gamma'}[s, o] = M_\gamma[s, o] \cup \{r\}$.

delete r from M[x, y] : when x is instantiated with s and y with o, this operation succeeds if and only if $s \in S_\gamma$ and $o \in O_\gamma$. The result is $M_{\gamma'}[s, o] = M_\gamma[s, o] - \{r\}$.

create subject x : when x is instantiated with s, this operation succeeds if and only if $s \in \mathcal{S} - S_\gamma$. The result is that $S_{\gamma'} = S_\gamma \cup \{s\}$, $O_{\gamma'} = O_\gamma \cup \{s\}$, $M_{\gamma'}[\,]$ has a row and column associated with s, and $M_{\gamma'}[s, o] = \emptyset$ for all $o \in O_{\gamma'}$.

create object y : when y is instantiated with o, this operation succeeds if and only if $o \in \mathcal{O} - O_\gamma$. The result is that $O_{\gamma'} = O_\gamma \cup \{o\}$, $M_{\gamma'}[\,]$ has a column associated with o, and $M_{\gamma'}[s, o] = \emptyset$ for all $s \in S_{\gamma'}$.

destroy subject x : when x is instantiated with s, this operation succeeds if and only if $s \in S_\gamma$. The result is that $S_{\gamma'} = S_\gamma - \{s\}$, and $O_{\gamma'} = O_\gamma - \{s\}$. $M_{\gamma'}[\,]$ has no row or column associated with s.

destroy object y : when *y* is instantiated with *o*, this operation succeeds if and only if $o \in O_\gamma$. The result is that $O'_\gamma = O_\gamma - \{o\}$, and $M_{\gamma'}[\ ]$ has no column associated with *o*.

LEAK SAFETY IN THE HRU SCHEME

Given an HRU system, we can easily check whether a subject has a right to an object in the current state by consulting the relevant cell in the access matrix. However, safety analysis, which asks questions regarding states that are reachable from the current state, is harder to solve. In [125], a system is considered to be safe for a given right if that right cannot be *leaked* to a cell, that is, be entered into a cell where it does not exist in the current state. Because this definition of safety utilizes the notion of the leakage of a right, we call this version of safety (r)-leak-safety.

Definition 4.3 Leak. Given a command α in an HRU system and the current state γ of the system, we say that α leaks the right *r* from γ if in the execution of α for some instantiation of its parameters in the state γ, the execution of a primitive operation (presumably of the form *enter r into* $M[s, o]$) in α enters *r* into a cell of the access matrix that did not contain *r* immediately before the execution of the primitive operation.

Definition 4.4 (r)-leak-safety. Given an HRU system with the start-state γ, a state-change rule ψ, and a right $r \in R_\psi$, we say that the system is (r)-leak-unsafe for *r* if and only if there is a state γ_n and a command α in the state-change rule ψ for the system such that:

1. $\gamma \stackrel{*}{\mapsto}_\psi \gamma_n$, and,

2. α leaks *r* from γ_n.

There are three subtleties about the above definition of leak-safety given in Definitions 4.3 and 4.4.

1. If a subject *s* has the right *r* over the object *o* in the start-state γ, and there is a sequence of state-changes by which *s* loses the right *r* to *o*, and then re-acquires it, the system is (r)-leak-unsafe for *r*.

   The HRU paper informally characterizes safety as: "…*whether, given some initial access matrix, there is some sequence of commands in which a particular generic right is entered in some place in the matrix where it did not exist **before***." Here the notion of "before" is not interpreted as the initial state but rather as the state immediately before the execution of one primitive operation.

2. When a right is entered and then deleted in the same command, the command is considered to "leak" the right, and the system is considered to be unsafe. This point is explicitly made in the HRU paper: "…*note that (a command)* α *leaks (the right) r from (the state) Q even if* α *deletes r after entering it*." The rationale provided for this is that there may be "code" in between

the entering of the right and the deletion of the right that is not "directly modeled". In other words, the HRU paper assumes that the execution of a command is non-transactional, and transient states during the execution of a command is visible.

3. A command may leak a right even when the leaky instantiation of command fails to complete its execution. All that is required is there is some operation in the "leaky" command that enters the right, where it does not exist in the access matrix, immediately preceding the execution of the operation. Even if a subsequent primitive operation fails so that one cannot apply the command as a whole to reach a new state, the system is considered to be (r)-leak-unsafe. This also shows that the execution of a command is assumed to be non-transactional.

Note that the non-transactional nature of command execution is only allowed for the last command in a sequence of commands that leaks a right. For example, consider a sequence of two commands such that during the execution of the first command (after some primitive operations in the command have been completed, but before all primitive operations in the command are completed), one executes a second command, which leads to the entering of a right r. This is not considered unsafe according to Definition 4.4, since one cannot apply a command to a transient state during the execution of the first command. This shows that regarding the question of whether a command execution is transactional or not, the HRU definitions of safety use inconsistent assumptions.

We have discussed some both explicit and implicit design choices made in the HRU notion of safety. Whether these choices are reasonable depends on whether they accurately model real access-control systems. It is unclear what kinds of systems are exhibits inconsistent behavior for command executions, as assumed in the HRU paper. We will explore other possible definition of safety later.

MAIN RESULTS ABOUT LEAK SAFETY IN THE HRU SCHEME

The main results about (r)-leak-safety are as follows:

- **Determining whether an HRU system is (r)-leak-unsafe is undecidable in general [125]].** The proof is by a reduction from the Turing Machine halting problem. The basic idea is that one can use an access matrix to simulate the tape of a Turing machine and use one command to simulate each state transition tuple in the Turing machine.

- **Determining whether a bi-conditional HRU system is (r)-leak-unsafe is undecidable [124].** A bi-conditional HRU system is one in which all commands are bi-conditional, i.e., each command contains at most two conditions. The proof is by a reduction from the Post Correspondence Problem.

- **Determining whether a mono-operational HRU system is (r)-leak-unsafe is NP-complete [125].** A mono-operational HRU system is one in which all commands have exactly one primitive operation in its body. The key observations that the problem is in NP are as follows. Because each command is mono-operational, any command that creates a new object cannot mark the new object (by adding some rights to the row or column corresponding to the object) after creating it. Therefore, any two newly created objects are equivalent. Furthermore, the HRU commands check only the presence of rights (but not the absence of rights) in the preconditions. Together, these mean that for a sequence of HRU commands that results in a leak, one can obtain a sequence that also results in a leak by doing the following: (1) remove all commands that delete rights or destroy subjects or objects; (2) remove all commands that create new subjects and objects except the first one (and change it to a create-subject command if it is a create-object command) and replace all later occurrences of new subjects or objects with the single new subject; and (3) remove commands that add a right in a cell that already has it. The resulting sequence has length at most polynomial in the size of the input, and can be verified to indeed cause a leak in time linear in the size of the sequence. In summary, if a mono-operational HRU system is unsafe, then there exists a polynomial-size witness that can be verified efficiently.

OTHER VERSIONS OF SAFETY

It is worthwhile to point out that (r)-leak-safety is not the only meaningful notion of safety. Rather than asking whether a right r can be leaked anywhere, one may be interested in knowing whether a right r about a specific object o can be leaked or not; we call this (r,o)-leak-safety. Similarly, one may be interested in knowing whether a right r about a specific object o can be leaked to a specific subject s; we call this (s,r,o)-leak-safety. The three complexity results about (r)-leak-safety also hold for (r,o)-leak-safety and (s,r,o)-leak-safety.

As discussed above, the definition of leak-safety relies on inconsistent interpretation of whether command executions are transactional. It seems more natural to assume that command executions are transactional. Also, it seems counter-intuitive to treat the entering of a right that already exists in the initial state as unsafe. Taking these into consideration, we arrive at the following notion of safety.

Definition 4.5 (r)-simple-safety. Given an HRU system with the start-state γ, a state-change rule ψ, and a right r in the system, we say that the system is (r)-simple-unsafe if and only if there is a state γ_n such that the following conditions hold:

1. The state γ_n is reachable, i.e., $\gamma \overset{*}{\mapsto}_\psi \gamma_n$.

2. In state γ_n, the right r exists in a cell where it does not exist in the state-state γ, i.e., there exist $s \in S_{\gamma_n}$ and $o \in O_{\gamma_n}$ such that first $r \in M_{\gamma_n}[s, o]$, and furthermore, either $s \notin S_\gamma$, or $o \notin O_\gamma$, or $r \notin M_\gamma[s, o]$.

One can also define (r,o)-simple-safety and (s,r,o)-simple-safety. The same complexity results apply to simple safety, as the proofs do not rely on whether the last command is transactional or not.

The fact that safety analysis is undecidable in the HRU scheme motivated many researchers to design access-control schemes in which safety analysis can be efficiently decided. For example, Jones et al. introduced the Take-Grant scheme [144; 166], in which safety can be decided in linear time. Sandhu et al. introduced the Schematic Protection Model [226], the Extended Schematic Protection Model [19; 20], and the Typed Access Matrix model [228]. These access-control schemes, however, are not used in the practice. The notions of safety studied in these later papers are simple safety rather than leak safety.

4.2 RBAC

Rather than inventing new access-control schemes for the sake of making safety analysis decidable, we believe it is more fruitful to study the safety analysis problem for access-control schemes that are used in practice. We now study analysis problems in Role-Based Access Control (RBAC) [23; 232]. In RBAC, there are users, roles, and permissions. Roles in RBAC form a middle layer between users and permissions, thus simplifying the management of the many-to-many relationships between users and permissions. Permissions are associated with roles, and users are granted membership in appropriate roles, thereby acquiring the roles' permissions. More advanced RBAC models include role hierarchies. A role hierarchy extends the role layer to have multiple layers, thereby further reducing the number of relationships that must be managed.

RBAC has been very widely used in the commercial world. Most major information technology vendors offer products that incorporate some form of RBAC. The commercial success of RBAC can be attributed to the following two advantages it has over previous approaches: reduced cost of administration and the ability to configure and enforce commercial security policies.

Large RBAC systems may have hundreds of roles and tens of thousands of users. For example, a case study carried out with Dresdner Bank, a major European bank, resulted in an RBAC system that has about 40,000 users and 1,300 roles [234]. In RBAC systems of this size, administration has to be decentralized, as it is impossible for one system security officer (SSO) or a small set of completely trusted administrators to administer the entire system.

Existing schemes for administering RBAC [81; 82; 203; 230; 231; 233] use delegation to decentralize the administration of RBAC to partially trusted users. While the use of delegation in administration of RBAC systems greatly enhances flexibility and scalability, it may also reduce the organization's control over its resources and jeopardize the persistence of security properties such as safety and availability. One major advantage that RBAC has over Discretionary Access Control (DAC) [6; 120] is that when using RBAC an organization (represented by the SSO in the system) has central control over its resources. This is different from DAC, in which the creator of a resource determines who can access the resource. In most organizations, even when a resource is created by an employee, often the resource is still owned by the organization, and the organization wants some level of control over how the resource is to be shared. Security analysis techniques can be used to ensure the preservation of desired security properties while delegating administration privileges.

We now define the access-control scheme that we study here, the URA97 RBAC scheme, which is based on the ARBAC97 administrative scheme for RBAC [230; 231]. URA97 is one of the three components of ARBAC97 [231]. The other components of ARBAC97 are PRA97 and RRA97, for administering permission-role assignment/revocation and the role hierarchy, respectively. Here, we study the effect of decentralizing user-role assignment and revocation, and assume that changes to the permission-role assignment relation and the role hierarchy are centralized, i.e, made only by trusted users. In other words, whoever is allowed to make changes to permission-role assignment and the role hierarchy will use security analysis, and they will only make those changes that do not violate desirable security properties.

We assume that there are three countable sets: $\mathcal{U}$ (the set of all possible users), $\mathcal{R}$ (the set of all possible roles), and $\mathcal{P}$ (the set of all possible permissions). While the set of all users in any RBAC state is finite, the set of all users that could be added is potentially unbounded. One can think of $\mathcal{U}$ as the set of all possible user-identifiers in a system.

Recall that an access-control scheme can be specified by giving $\langle \Gamma, Q, \vdash, \Psi \rangle$. We not specify them for RBAC.

States (Γ).   An RBAC state γ is a 6-tuple $\langle UA, PA, RH, CA, CR, CO \rangle$. We call UA, PA, and RH parts of the *basic state*, and CA, CR, CO parts of the *administrative state*. The basic state is described below; the administrative state is described when we discuss state transitions.

The user assignment relation $UA \subseteq \mathcal{U} \times \mathcal{R}$ associates users with roles, the permission assignment relation $PA \subseteq \mathcal{R} \times \mathcal{P}$ associates roles with permissions, and the role hierarchy relation $RH \subseteq \mathcal{R} \times \mathcal{R}$ is an irreflexive and acyclic relation over $\mathcal{R}$. We use $\succeq_{RH}$ to denote the partial order induced by RH, i.e., the transitive and reflexive closure of RH. That $r_1 \succeq_{RH} r_2$ means that every user who is authorized for r_1 is also authorized for r_2 and every permission that is associated with r_2 is also associated with r_1.

Given a state γ, each user has a set of roles for which the user is authorized. We formalize this by defining for every state γ a function authorizedRoles : $\mathcal{U} \to 2^{\mathcal{R}}$

$$\text{authorizedRoles}(u) = \{\, r \in \mathcal{R} \mid \exists r_1 \in \mathcal{R} \,[(u, r_1) \in UA \wedge (r_1 \succeq_{RH} r)]\,\}$$

When $r \in$ authorizedRoles(u), we say that the user u is authorized for the role r, or equivalently, u is a member of r. We also define *down*(r) to be the set of all roles dominated by r and *up*(r) to be the set all roles that dominate r as follows:

$$down(r) = \{\, r' \in \mathcal{R} \mid r \succeq_{RH} r' \,\} \qquad up(r) = \{\, r' \in \mathcal{R} \mid r' \succeq_{RH} r \,\}$$

State Transition Ψ.   There are two kinds of actions that may change a state in the URA97 scheme: assignments and revocations. Whether these actions succeed or not when applied in a state depends on the administrative state of γ, namely CA, CR, and CO, which we describe below.

1. The relation $CA \subseteq \mathcal{R} \times C \times 2^{\mathcal{R}}$ determines who can assign users to roles and the preconditions these users must satisfy. C is the set of conditions, which are expressions formed using

roles, the binary operators $\cap$ and $\cup$, the unary operator $\neg$, and parentheses. A tuple $\langle r_a, c, rset \rangle$ in CA means that members of the role r_a can assign any user whose role memberships satisfy the condition c, to any role $r \in rset$. For example, $\langle r_0, r_1 \cap r_2 \cap \neg r_3, \{r_4\} \rangle \in CA$ means that a user that is a member of the role r_0 is allowed to assign a user that is a member of both r_1 and r_2 but is not a member of r_3, to be a member of r_4.

2. The relation $CR \subseteq \mathcal{R} \times 2^{\mathcal{R}}$ determines who can remove users from roles. That $\langle r_a, rset \rangle \in CR$ means that the members of role r_a can remove a user from a role $r \in rset$. Unlike the relation CA, there is no preconditions in the relation CR defined in URA97 [231].

   We assume that CA and CR satisfy the property that the administrative roles are not affected by CA and CR. The administrative roles are those that appear in the first component of each tuple in CA or CR. These roles should not appear in the last component of any CA or CR tuple. This condition is satisfied in URA97, which assumes the existence of a set of administrative roles that is disjoint from the set of normal roles.

3. CO is a set of mutually exclusive role constraints. Each constraint in CO has the form smer$\langle \{r_1, \ldots, r_m\}, t \rangle$ where smer is keyword and stands for statically mutually exclusive roles, each r_i is a role, and m and t are integers such that $1 < t \le m$. This constraint forbids a user from being a member of t or more roles in $\{r_1, \ldots, r_m\}$. We say that a set R of roles satisfies a constraint smer$\langle \{r_1, \ldots, r_m\}, t \rangle$ if and only if $|R \cap \{r_1, \ldots, r_m\}| < t$, where $| |$ gives the cardinality of a set.

   For example, smer$\langle \{r_1, r_2\}, 2 \rangle$ means that no user is allowed to be a member of both r_1 and r_2. In an RBAC state γ, if $r_1 \in$ authorizedRoles(u) for a user u, then an assignment action that assigns the user u to any role in $up(r_2)$ would fail because of the constraint.

Ψ consists of a single state-transition rule, ψ, which is a set of actions:

$$\psi = \{assign(u_a, u_t, r_t) \mid u_a, u_t \in \mathcal{U} \land r_t \in \mathcal{R}\}$$
$$\cup \{revoke(u_a, u_t, r_t) \mid u_a, u_t \in \mathcal{U} \land r_t \in \mathcal{R}\}$$

1. An assignment action $assign(u_a, u_t, r_t)$ means that the user u_a assigns the user u_t to the role r_t. When this action is applied to an RBAC state γ, it succeeds if and only if the following three conditions hold:

   (a) $(u_t, r_t) \notin UA$, i.e., the user u_t is not already assigned the role r_t.

   (b) There exists a tuple $\langle r_a, c, rset \rangle \in CA$ such that $r_a \in$ authorizedRoles(u_a), authorizedRoles(u_t) satisfies c, and $r_t \in rset$.

   (c) authorizedRoles$(u_t) \cup down(r_t)$ satisfies every constraint in CO, i.e., the new role memberships of u_t do not violate any constraint.

When the assignment action is successfully applied to an RBAC state γ, the resulting state γ' differs from γ only in the user-role relation. The result of a successful application is $UA' = UA \cup \{(u_t, r_t)\}$. When the application is not successful, the state does not change.

2. A revocation action is of the form $revoke(u_a, u_t, r_t)$, which means that the user u_a revokes the user u_t from the role r_t. When this action is applied to an RBAC state γ, it succeeds if and only if the following two conditions hold:

   (a) $(u_t, r_t) \in UA$, i.e., the user u_t is assigned to the role r_t.

   (b) There exists a tuple $\langle r_a, rset \rangle \in CR$ such that $r_a \in$ authorizedRoles(u_a), and $r_t \in rset$.

   When the revocation action is successfully applied to an RBAC state γ, the resulting state γ' differs from γ only in the user-role relation. The result of a successful application is $UA' = UA - \{(u_t, r_t)\}$. When the application is not successful, the state does not change.

SECURITY ANALYSIS PROBLEMS IN URA97

We now define a Security Analysis Problem (SAP) in the URA97 RBAC scheme.

Definition 4.6   URA-SAP. A URA-SAP instance is given by an RBAC state $\gamma = \langle UA, PA, RH, CA, CR, CO \rangle$, a set $A_T \subseteq \mathcal{U}$ of trusted users, and a query.

We deliberately leave the syntax for queries unspecified in the above definition. Different kinds of queries may be needed for different policy analyses. The simplest kind is to ask whether a user u is a member of a role r. More sophisticated queries may ask whether a user's role membership satisfy a condition, e.g., $r_1 \cup (r_2 \cap \neg r_3)$, or whether the set of members of one role is a subset of the set of members of another role.

An important observation is that the simplest query which asks whether a user is a member of a role can be used to handle several other kinds of queries. For example, if one wants to know whether the system can reach a state in which u's role membership includes a set $\{r_1, r_2\}$ and excludes $\{r_3\}$, one can add a new user u_a, two new roles r_a and r_t, a user assignment (u_a, r_a), and a new tuple $(r_a, (r_1 \cap r_2 \cap \neg r_3), \{r_t\})$ to CA, and asks whether the new system can reach a state in which $u \in r_t$ is true. The new system can reach a state in which $u \in r_t$ if and only it can reach a state in which u satisfies $(r_1 \cap r_2 \cap \neg r_3)$, which happens if and only if the original system can reach a state in which u satisfies $(r_1 \cap r_2 \cap \neg r_3)$. Similarly, if one wants to know whether the system can reach a state in which u possesses a certain set of permissions, one can compute the role condition that is necessary and sufficient to have the permissions and then translate that into a query about a single role.

Definition 4.7   URA-RC-SAP. A URA-RC-SAP instance is a special case of URA-SAP in which a query has the form $u \in r$.

COMPUTATIONAL COMPLEXITY

We now study the computational complexity of URA-SAP. In particular, we show that URA-RC-SAP is **PSPACE**-complete. The main source of the complexity of SAP is that the state space that needs to be explored is potentially large. We would like to understand how different features in URA97 affect this search space; therefore, we consider special cases of URA-SAP that result from restricting the URA scheme in various ways. Answers to the following questions affect the computational complexity of URA-SAP.

1. *What queries are considered?* If queries are allowed to contain conjunctions and disjunctions of roles, then URA-SAP is likely to be intractable. For example, in [163], one can pose a query that asks whether the set of users who satisfy $((r_1 \cup r_2) \cap r_3)$ is always a subset of the set of users that satisfy $((r_1 \cup r_2) \cap (r_2 \cup r_3))$. These sophisticated queries can encode propositional formulas, resulting in **NP**-hardness.

2. *Do the preconditions involve only conjunctions?* Each tuple in *CA* has a precondition. It is conceivable that if the precondition involves arbitrary conjunction, disjunction, and negation of roles, then this could make the problem intractable; however, such a result would be less insightful and of less practical interest. In practical systems, one would not expect the precondition to be a very complicated logical formula. Hence one may want to consider the special case in which each precondition is a conjunction of roles or their negations.

3. *Is negation allowed in preconditions in CA?* When preconditions in *CA* may contain negation, one needs to consider the revocation of a users' role memberships in order to satisfy the precondition and be assigned to a new role.

4. *Are role mutual exclusion constraints allowed, i.e., is CO = { }?* When constraints are allowed, one may need to consider revocations in order to assign a user to a new role.

5. *Are revocations allowed, i.e., whether CR = { }?* One may want to consider the special case that role memberships cannot be revoked.

We now summarize the variations in Figure 4.1. The computational complexity for these variations are stated in the following theorem. These results are also summarized in Figure 4.2. Proofs for these results can be found in [140].

Theorem 4.8   *The computational complexity for URA-RC-SAP and its various subcases are as follows.*

- URA-RC-SAP is **PSPACE**-complete.

- URA-RC-SAP[*CA* (conjunctive), *CR*, *CO*] is **PSPACE**-complete.

- URA-RC-SAP[*CA* (positive conjunctive), *CR*, *CO*] is **PSPACE**-complete.

- URA-RC-SAP[*CA* (conjunctive), *CR*, *CO* = { }] is **PSPACE**-complete.

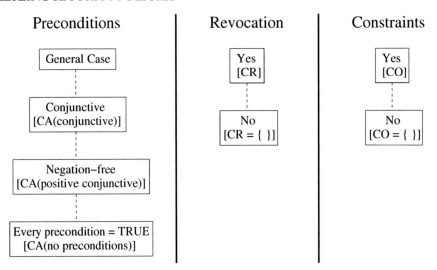

Figure 4.1: The possible variations in the features we consider for the preconditions in *CA*, revocation and constraints. A dotted line connects a case with a subcase. For example, negation-free preconditions ([*CA* (positive conjunctive)]) is a subcase of conjunctive preconditions ([*CA* (conjunctive)]). Various combinations from the three columns are possible. For example, we can consider the analysis problem with negation-free preconditions, with revocation, but without constraints, which corresponds to URA-RC-SAP[*CA* (positive conjunctive), *CR*, *CO* = { }].

- URA-RC-SAP[*CA* (conjunctive), *CR* = { }, *CO*] is **NP**-complete.

- URA-RC-SAP[*CA* (conjunctive), *CR* = { }, *CO* = { }] is **NP**-complete.

- URA-RC-SAP[*CA* (positive conjunctive), *CR* = { }, *CO*] is **NP**-complete.

- URA-RC-SAP[*CA* (positive conjunctive), *CR*, *CO* = { }] is in **P**.

- URA-RC-SAP[*CA* (no preconditions), *CR*, *CO*] is in **P**.

4.3   SECURITY POLICIES

4.3.1   TRUST MANAGEMENT

Systems with shared computing resources use access-control mechanisms for protection. The main issues in access control of shared computing resources are *authentication*, *authorization*, and *enforcement*. Identification of principals is handled by authentication. Authorization addresses the question

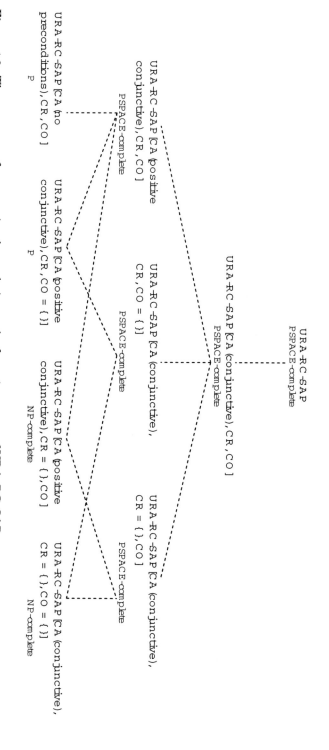

Figure 4.2: The summary of computational complexity results for various cases of URA-RC-SAP.

"Should a request *r* by a specific principal *A* be allowed?" Enforcement addresses the problem of restricting the system to perform only authorized operations during an execution.

In a centralized system, authorization is based on the *closed-world assumption*, i.e., all of the parties are known and trusted. In a distributed system, where not all the parties are known *a priori*, the closed-world assumption is not appropriate.

Trust-management systems [42; 44; 254] solve the authorization problem for shared computing resources by defining a formal language for expressing security policies, and rely on an algorithm to determine when a specific request is allowable. A request is a triple consisting of the identity of the entity (usually a public key), an access request, and a proof of authorization (usually a chain of signed assertions). The problem of creating a proof of authorization given the security policy, the identity of a principal, and an access request is called *certificate-chain discovery*. The problem of determining whether a purported proof of authorization complies with a security policy is called *compliance checking*. (See Fig. 4.3.)

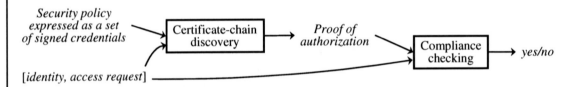

Figure 4.3: Trust management.

Several trust management systems, such as Binder [93], Keynote [42], Referee [67], and SPKI/SDSI [104; 106], have been proposed. A survey of trust management systems, along with a formal framework for understanding them, is presented in [254].

The following concrete scenarios motivate the technical problems addressed in this section:

Distributed authorization. Suppose that there are two sites, Bio and CS, which correspond to the biology and the computer science departments, respectively. Suppose that two professors, *Alice* from CS and *Bob* from Bio, are collaborating on a project in which several of *Alice's* students are involved, and that *Bob* wants to provide access to a server *V* to all of *Alice's* students.

Using traditional mechanisms, *Bob* will have to know the identity of all of *Alice's* students. Ideally, however, *Bob* should be able to assert the statement given below, and authorization should happen seamlessly:

Server V is accessible to all of Alice's students.

In other words, the management of the group of *Alice's* students should be performed by *Alice*; to grant authorization, *Bob* should not be required to know the identity of the group members. This also has the advantage of decentralized management, i.e., if the group changes (for example, a student leaves), the authorization algorithm should seamlessly incorporate this new information without *Bob* having to explicitly change access-control lists (ACLs) on server *V*.

Privacy-preserving authorization. Suppose that company X offers insurance to cover prescription-drug expenses that are not covered by a patient's health-maintenance organization (HMO). (For example, the HMO might have a very high deductible for drugs, which would be covered by the additional insurance.) However, suppose that company X only wants to offer this product to patients of a certain hospital H. For Charles to be able to buy such insurance, he needs to prove that he is a patient of hospital H. Ideally, Charles should be able to authorize himself to X without revealing sensitive information, e.g., that he is a patient of H's AIDS clinic.

Suppose that there are two certificate chains that prove that Charles is a patient of H, where one reveals that Charles is a patient in the internal-medicine clinic and the other reveals that Charles is a patient in the AIDS clinic. For obvious reasons, Charles would prefer to use the former certificate chain. In other words, Charles is interested in the answer to the question

> *What is the certificate chain that reveals the least amount of information about me?*

Such privacy-related issues can be addressed in the framework that will be presented below.

Analysis of access-control policies. Returning to the example of server V and access rights for Alice's students, either professor might be interested in knowing whether service will continue uninterrupted over the weekend:

> *Will Alice's students be prevented from accessing server V after the certificates due to expire this weekend have expired?*

A student of Alice may be interested in knowing how long he has before he is unable to access server V (e.g., to know how long he can procrastinate before he must complete his project, which requires access to server V). He is interested in the answer to the question

> *What is the maximum expiration value among all certificate chains that provide me with access to server V?*

4.3.2   SPKI/SDSI

SPKI/SDSI [69; 104; 106] is a trust-management framework that is distinguished by its support for local name spaces and delegation:

Local name spaces: Any party can define local names and specify what those names mean: "name certificates" define the names available in an issuer's local name space.

Delegation: "Authorization certificates" either grant authorizations directly, or grant authorization indirectly by delegating the ability to grant authorizations.

All certificates employ a uniform mechanism for declaring names; how a name is resolved depends on the certificates that define the local name spaces.

4.3.2.1 Principals and Names

In SPKI/SDSI, all *principals* are represented by their public keys, i.e., the principal *is* its public key. A principal can be an individual, process, host, or any other entity. $\mathcal{K}$ denotes the set of public keys. Specific keys are denoted by K, K_A, K_B, K', etc. An *identifier* is a word over some alphabet Σ. The set of identifiers is denoted by $\mathcal{A}$. Identifiers will be written in typewriter font, e.g., A and Bob.

A *term* is a key followed by zero or more identifiers. Terms are either keys, local names, or extended names. A *local name* is of the form K A, where $K \in \mathcal{K}$ and A $\in \mathcal{A}$. For example, K Bob is a local name. Local names are important in SPKI/SDSI because they create a decentralized name space. The local name space of K is the set of local names of the form K A. An *extended name* is of the form $K \sigma$, where $K \in \mathcal{K}$ and σ is a sequence of identifiers of length greater than one. For example, K UW CS faculty is an extended name.

4.3.2.2 Certificates

SPKI/SDSI has two types of certificates, or "certs". The first type of certificate, called a *name cert*, provides definitions of local names. Authorizations are specified using *authorization certs* (or *auth certs*, for short).

Name Certificates. A name cert provides a definition of a local name in the issuer's local name space. Only key K may issue or sign a cert that defines a name in its local name space. A name cert is a signed four-tuple (K, A, S, V):

- The issuer K is a public key and the certificate is signed by K.

- A is an identifier.

- The subject S is a term in $\mathcal{T}$. Intuitively, S gives additional meaning for the local name K A.

- The *validity specification* V provides information regarding the validity of the certificate. Usually, V takes the form of an interval $[t_1, t_2]$, i.e., the cert is valid from time t_1 to t_2, inclusive.

Authorization Certificates. An auth cert grants or delegates a specific authorization from an issuer to a subject. Specifically, an auth cert is a five-tuple (K, S, D, T, V), where

- The *issuer* K is a public key, which is also used to sign the cert. The issuer is the principal granting a specific authorization.

- The *subject* S is a term.

- If the *delegation bit* D is turned on, then a key that receives the authorization can delegate this authorization to other keys.

- The *authorization specification* T specifies the permission being granted. For example, it may specify a permission to read a specific file or a permission to login to a particular host. For instance, the authorization specification (dir /afs/cs.wisc.edu/public/tmp) refers to the resource /afs/cs.wisc.edu/public/tmp.[1]

---

[1] In SPKI/SDSI, authorization specifications are parenthesized lists in a Lisp-like s-expression notation.

We will assume that there is a partial order on authorization specifications, e.g., a partial order $T \supseteq T'$ might correspond to the fact that T is "more permissive" than T'. In the example shown below, T is more permissive than T' (i.e., the set of actions defined by T are a strict superset of the set of actions defined by T'.[2]

T': ((dir /afs/cs.wisc.edu/public/tmp) read)
T: ((dir /afs/cs.wisc.edu/public/tmp) (* set read write))

- The *validity specification* V for an auth cert is the same as in the case of a name cert.

4.3.2.3 Certificates as Rewrite Rules

A *labeled rewrite rule* is a triple $L \xrightarrow{T} R$, where L and R are terms and T is an authorization specification. Authorization specifications are equipped with a lattice structure; following the literature, the lattice operations will be denoted by $\cap$ and $\cup$, respectively.[3] $\hat{T}$ is the authorization specification such that for all other authorization specifications t, $\hat{T} \cap t = t$, and $\hat{T} \cup t = \hat{T}$ (i.e., $\hat{T}$ is the top element of the lattice.) Often we will write $\xrightarrow{\hat{T}}$ simply as $\longrightarrow$, i.e., a rewrite rule of the form $L \longrightarrow R$ has an implicit label of $\hat{T}$.

We will treat certs as labeled rewrite rules.

- A name cert (K, A, S, V) will be written as a labeled rewrite rule $K\,A \xrightarrow{\hat{T}} S$.

- An auth cert (K, S, D, T, V) will be written as $K\,\square \xrightarrow{T} S\,\square$ if the delegation bit D is turned on; otherwise, it will be written as $K\,\square \xrightarrow{T} S\,\blacksquare$.

By convention, in authorization problems we only consider valid certificates; hence, the validity specification V for a certificate does not appear as part of its rewrite rule.[4] In practice, as a preprocessing step we first check the validity specification V of each certificate in use and discard those that are invalid. Henceforth, we assume that only valid certificates are considered during certificate-chain discovery.

Because we only use labeled rewrite rules, we refer to them as rewrite rules, or simply rules. A term S appearing in a rule can be viewed as a string over the alphabet $\mathcal{K} \cup \mathcal{A}$, in which elements of $\mathcal{K}$ appear only in the beginning. For uniformity, we also refer to strings of the form $S\,\square$ and $S\,\blacksquare$ as terms.

Assume that we are given a labeled rewrite rule $L \xrightarrow{T} R$ that corresponds to a cert. Consider a term $S = LX$. In this case, the labeled rewrite rule $L \xrightarrow{T} R$ applied to the term S (denoted by $(L \xrightarrow{T} R)(S)$) yields the term RX. Therefore, a rule can be viewed as a function from terms to terms that rewrites the left prefix of its argument, for example,

$$(K_A\,\text{Bob} \longrightarrow K_B)(K_A\,\text{Bob}\,\texttt{myFriends}) = K_B\,\texttt{myFriends}$$

[2]"(* set read write)" denotes read/write permission; "read" denotes read permission.
[3]Intersection and union of authorization specifications are discussed in detail in [106; 137].
[4]For certain generalized authorization problems, V will be used to derive a weight for the corresponding rewrite rule (see §4.3.5.1).

Consider two rules $c_1 = (L_1 \xrightarrow{T} R_1)$ and $c_2 = (L_2 \xrightarrow{T'} R_2)$, and, in addition, assume that L_2 is a prefix of R_1, i.e., there exists an X such that $R_1 = L_2 X$. Then the *composition* $c_2 \circ c_1$ is the rule $L_1 \xrightarrow{T \cap T'} R_2 X$.[5] For example, consider the two rules:

$$c_1: \quad K_A \text{ friends} \xrightarrow{T} K_A \text{ Bob myFriends}$$
$$c_2: \quad K_A \text{ Bob} \xrightarrow{T'} K_B$$

The composition $c_2 \circ c_1$ is $K_A \text{ friends} \xrightarrow{T \cap T'} K_B \text{ myFriends}$. Two rules c_1 and c_2 are called *compatible* if their composition $c_2 \circ c_1$ is well-defined.[6]

A *certificate chain* ch is a sequence of certificates $\langle c_1, c_2, \cdots, c_k \rangle$. The label of a certificate chain $ch = \langle c_1, c_2, \cdots, c_k \rangle$, denoted by $L(ch)$, is the label obtained from $c_k \circ c_{k-1} \circ \cdots \circ c_1$.

4.3.2.4 The Authorization Problem in SPKI/SDSI

In traditional discretionary access control, each protected resource has an associated access-control list, or ACL, describing which principals have various permissions to access the resource. An auth cert (K, S, D, T, V)—which would normally be associated with a certain resource R that is owned by principal K—can be viewed as an ACL entry for R, where keys (principals) represented by the subject S are given permission to access R with right T. However, SPKI/SDSI goes beyond traditional ACLs through its support for local name spaces and delegation.

Example 4.3.1 Using the scenario introduced earlier, we can explain how distributed groups are supported by the SPKI/SDSI framework. Recall that *Bob* and *Alice* are faculty members in two different departments, and are collaborating on a project, along with several of *Alice's* students. Suppose that *Bob* wants to delegate to *Alice* execute rights on a server V. In addition, suppose that *Alice* wants to delegate those same rights on V to her students but not allow them to delegate those rights any further.

We assume that each resource R has a *unique owner* (denoted by $K_{owner[R]}$). For instance, in the example shown in Fig. 4.4, which is discussed below, K_{Bob} is the owner of the resource V, and $\{[V, \text{execute}]\}$ denotes execute rights on V.

There are three components to a SPKI/SDSI authorization scenario:

Certificate issuance. With SPKI/SDSI, an access-control policy is expressed as a collection of certificates. In particular, each user can issue various auth and/or name certs. For instance, the owner of a resource R might issue an auth cert $(K_{owner[R]}, S, D, T, V)$ to declare that principals represented by subject S can exercise right T on R.

---

[5]Our notation differs from that used in [69]. For rule application, we write $(L \rightarrow R)(S)$ instead of $S \circ (L \rightarrow R)$. For the composition of $C_1 = (L_1 \rightarrow L_2 X)$ and $C_2 = L_2 \rightarrow R_2$, we write $C_2 \circ C_1$ instead of $C_1 \circ C_2$ (so that $(C_2 \circ C_1)(L_1) = (C_2(C_1(L_1))) = R_2 X$).

[6]In general, the composition operator $\circ$ is not associative. In particular, c_3 can be compatible with $c_2 \circ c_1$, but c_3 might not be compatible with c_2. Therefore, $c_3 \circ (c_2 \circ c_1)$ can exist when $(c_3 \circ c_2) \circ c_1$ does not exist. However, when $(c_3 \circ c_2) \circ c_1$ exists, so does $c_3 \circ (c_2 \circ c_1)$; moreover, the expressions are equal when both are defined. Thus, we allow ourselves to omit parentheses and assume that $\circ$ is right associative.

	Certificates	Labeled Rewrite Rules
(1)	$(K_{Bob}, K_{Alice}, 1, \{[V, execute]\}, [t_1, t_2])$	$K_{Bob} \xrightarrow{\{[V, execute]\}} K_{Alice}\ \square$
(2)	$(K_{Alice}, K_{Alice}\ student, 0, \{[V, execute]\}, [t_1, t_2])$	$K_{Alice} \xrightarrow{\{[V, execute]\}} K_{Alice}\ student\ \blacksquare$
(3)	$(K_{Alice}, student, K_X, [t_1, t_2])$	$K_{Alice}\ student \longrightarrow K_X$
(4)	$(K_{Alice}, student, K_Y, [t_1, t_2])$	$K_{Alice}\ student \longrightarrow K_Y$
(5)	$(K_{Alice}, student, K_Z, [t_1, t_2])$	$K_{Alice}\ student \longrightarrow K_Z$

Figure 4.4: SPKI/SDSI certificates and their corresponding labeled rewrite rules.

Column 2 of Fig. 4.4 shows the certificates issued. K_{Bob} and K_{Alice} are public keys for *Bob* and *Alice*. The public keys for *Alice's* students X, Y, and Z are K_X, K_Y, and K_Z, respectively. Certificate (1) is an auth cert that is signed by *Bob*, and states that *Alice* has the right to execute on server V and can delegate that right (indicated by the 1 in the third field). The fifth field indicates that the certificate is valid in the time interval $[t_1, t_2]$. Certificates (2), (3), (4), and (5) are certificates that are signed by *Alice*. In auth cert (2), *Alice* grants to her students the right to execute on server V, but this right cannot be further delegated (indicated by the 0 in the third field). Name certs (3), (4), and (5) state that X, Y, and Z are *Alice's* students.

Certificate-chain discovery. To be able to exercise a right T_R on a resource R, a user U first performs certificate-chain discovery to obtain a proof that he has sufficient authorization to exercise T_R on R. This can be achieved by executing a certificate-chain-discovery algorithm, and if the algorithm finds that U is authorized, it returns a proof in the form of a finite set of certificate chains $\{ch_1, \ldots, ch_m\}$.

For example, assume that student X wants to execute a job on server V; for this, he needs to access V, exercising the authorization $\{[V, \textit{execute}]\}$ on V. Assuming that certificates (1), (2), and (3) of Fig. 4.4 are all valid, certificate-chain discovery would return the singleton set of chains $\{ch_1\}$, where $ch_1 = \langle (1), (2), (3) \rangle$.

Compliance checking. In general, after a user U obtains a set of certificate chains $SCH = \{ch_1, \ldots, ch_m\}$ from certificate-chain discovery, he presents SCH to $K_{\text{owner}[R]}$, the owner of the resource R to which T_R refers. The owner grants permission for K_U to access R iff

(i) For each chain $ch_i = \langle c_1, c_2, \cdots, c_k \rangle \in SCH$, $(c_k \circ \cdots \circ c_2 \circ c_1)(K_{\text{owner}[R]}) \,\square\, \in \{K_U \,\square, K_U \,\blacksquare\}$

(ii) $T_R \subseteq \bigcup_{i=1}^{m} L(ch_i)$

In other words, the certificate chains that grant U access to resource R all correspond to paths that start at $K_{\text{owner}[R]}$ and reach either $K_U \,\square$ or $K_U \,\blacksquare$; moreover, the union ($\cup$) of the composed labels on the paths is a superset of T_R.

In our example, after making the certificate-chain-discovery request, "Does $K_{Bob} \,\square$ resolve to either $K_X \,\square$ or $K_X \,\blacksquare$ with authorization specification $\{[V, \textit{execute}]\}$?", student X presents $\{ch_1\} = \{\langle (1), (2), (3) \rangle\}$ to K_{Bob}, which performs checks (i) and (ii), and determines that

(i) The certificate chain $\langle (1), (2), (3) \rangle$ is acceptable because

- Certificate (1) gives the delegable right $\{[V, \textit{execute}]\}$ to Alice.
- Certificate (2) provides each of Alice's students with the non-delegable right $\{[V, \textit{execute}]\}$.
- Certificate (3) proves that X is a student of Alice.

(ii) $L(ch_1)$ is $\{[V, \textit{execute}]\}$, which equals the permissions that student X requires on server V.

Because this (singleton) set of certificate chains passes the compliance check, X will be permitted to execute his job on V.

Note that if Alice does not renew certificate (3) when it expires (say, because X left her group), the certificate chain $\langle(1), (2), (3)\rangle$ will no longer be valid. The fact that this can be done independently of Bob is what makes SPKI/SDSI a convenient mechanism for supporting groups that are (conceptually) *distributed*. $\square$

To sum up, the authorization problem addresses the question "Given a set of certs C and a request $r = (K_U, R, T_R)$, is K_U allowed to exercise authorization T_R on R?" A key issue for making this approach tenable is there must be a method for automatically discovering certificate chains (otherwise, users would have to find certificate chains by hand). A certificate-chain-discovery algorithm provides more than just a simple yes/no answer to the authorization question; in the case of a yes answer, it identifies a finite set of certificate chains that prove the result.

Formally, certificate-chain discovery attempts to find, after removing useless certificates, a finite set of certificate chains $\{ch_1, \ldots, ch_m\}$ such that

(i) For each chain $\quad ch_i = \langle c_1, c_2, \cdots, c_k \rangle \in SCH, \quad (c_k \circ \cdots \circ c_2 \circ c_1)(K_{\text{owner}[R]} \,\square\, \in \{K_U \,\square, K_U \,\blacksquare\}$

(ii) $T_R \subseteq \bigcup_{i=1}^{m} L(ch_i)$

An efficient certificate-chain-discovery algorithm for SPKI/SDSI was presented by Clarke et al. [69]. Jha and Reps [142] presented a different algorithm, based on the theory of pushdown systems. Jha and Reps also demonstrated how this translation enables many other questions to be answered about a security policy expressed as a set of certificates. Schwoon et al. [239] then introduced *weighted* PDSs, and showed how that formalism not only fixes a certain failing of previous work (§4.3.6.2), but also allows a variety of important security-policy issues to be addressed, such as privacy (§4.3.4), authorization expiration, accounting for recency of certificates, and finding a maximal-trust authorization (§4.3.5). The WPDS-based approach is surveyed in the next three sections.

4.3.3 THE BASIC CONNECTION BETWEEN SPKI/SDSI AND PUSHDOWN SYSTEMS

The following correspondence between SPKI/SDSI and pushdown systems was presented in [142]: let C be a (finite) set of certificates such that $\mathcal{K}_C$ and $\mathcal{I}_C$ are the keys and identifiers that appear in C, respectively; with C we associate the pushdown system $\mathcal{P}_C = (\mathcal{K}_C, \mathcal{I}_C \cup \{\square, \blacksquare\}, \Delta_C)$, i.e., the keys of C are the control locations and the identifiers form the stack alphabet; the rule set Δ_C is defined as follows:

- if C contains a name cert $K \text{ A} \longrightarrow K' \sigma$ (where σ is a sequence of identifiers), then Δ_C contains a rule $\langle K, \text{A} \rangle \hookrightarrow \langle K', \sigma \rangle$;

- if C contains an auth cert $K \,\square\, \longrightarrow K' \sigma \, b$ (where $b \in \{\square, \blacksquare\}$), then Δ_C contains a rule $\langle K, \square \rangle \hookrightarrow \langle K', \sigma b \rangle$.

For instance, consider the set of certificates C from column 2 of Fig. 4.4. The corresponding pushdown system $\mathcal{P}_C$ has the control locations $\{K_{Bob}, K_{Alice}, K_X, K_Y, K_Z\}$, the stack alphabet $\{\texttt{student}, \square, \blacksquare\}$, and the set of rules listed in column 3 of Fig. 4.4.

The usefulness of this correspondence stems from the following simple observation: A configuration $\langle K, \sigma \rangle$ of $\mathcal{P}_C$ can reach another configuration $\langle K', \sigma' \rangle$ if and only if C contains a chain of certificates that, when applied to $K\,\sigma$, yield $K'\,\sigma'$.

4.3.4   THE GENERALIZED AUTHORIZATION PROBLEM

§4.3.3 ignored authorization specifications that are part of SPKI/SDSI auth certs. In this section, we formally define the *generalized authorization problem*, or *GAP*, which account for SPKI/SDSI authorization specifications. Later, in §4.3.5, we show that several issues, such as validity, recency, and trust, can be formulated in the GAP framework.

We discuss how algorithms for solving reachability problems on WPDSs are a useful tool for solving problems related to certificate-chain discovery in SPKI/SDSI. The following correspondence between SPKI/SDSI and WPDSs was presented in [239]: let C be a (finite) set of certificates such that $\mathcal{K}_C$ and $\mathcal{I}_C$ are the keys and identifiers, respectively, that appear in C. Moreover, let $\mathcal{T}$ be the set from which the authorization specifications in C are drawn. Then $\mathcal{S}_C = (\mathcal{T}, \cup, \cap, \emptyset, \hat{T})$, where $\cap, \cup$ are the intersection and union of auth specs as discussed in [106; 137], forms an idempotent semiring with domain $\mathcal{T}$. We now associate with C the WPDS $\mathcal{W}_C = (\mathcal{P}_C, \mathcal{S}_C, f)$, where $\mathcal{P}_C$ is the PDS $(\mathcal{K}_C, \mathcal{I}_C \cup \{\square, \blacksquare\}, \Delta_C)$, as defined in §4.3.3, and f maps each rule to its corresponding authorization specification.

As noted in §4.3.3, a PDS configuration $\langle K, \sigma \rangle$ of $\mathcal{P}_C$ can reach another configuration $\langle K', \sigma' \rangle$ if and only if C contains a chain of certificates $\langle c_1, \ldots, c_k \rangle$ such that $(c_k \circ \cdots \circ c_1)(K\,\sigma) = K'\,\sigma'$. For a WPDS, the label of the certificate chain is precisely $L(\langle c_1 \cdots c_k \rangle)$. Thus, solving path problems on WPDSs (e.g., finding $\mathrm{JOVP}(S, T)$, for regular sets of configurations S and T) provides a way to find a set of certificate chains to prove that a certain principal K' is allowed to access a resource owned by principal K. Moreover, the solution of the problem identifies a set of certificate chains such that the union of their labels provides as large a set of rights as possible.

In the generalized-authorization framework that we now introduce, certificates are labeled with *weights* that are drawn from an arbitrary bounded idempotent semiring (see Defn. 2.2.6 of §2.2). As evidenced by Ex. 4.3.3 and the examples in §4.3.5, these weights need not be classical SPKI/SDSI authorization specifications [106; 137], but can concern other security-policy issues in trust management systems, such as privacy, validity, recency, and trust.

The choice of what weights to use depends on the specific issue being addressed, but the weights can often be thought of as a kind of metric on certificates—where the metric is extended (i) from certificates to certificate chains by means of the semiring's extend operator ($\otimes$), and (ii) from certificate chains to sets of certificate chains by means of the semiring's combine operator ($\oplus$).

In a *generalized authorization problem* (GAP), we are given a principal K, a set of name and auth certificates C, a resource R, and a weight on each certificate. The question that GAP addresses is essentially the same before—i.e., given $\mathcal{C}$, is K authorized to access resource R?—however, an authorization proof that solves a GAP has the least weight (in the semiring ordering) over all possible proofs. (Recall that with a bounded idempotent semiring $(D, \oplus, \otimes, \bar{0}, \bar{1})$, the structure $(D, \oplus, \bar{0})$ is a join-semilattice, where $\bar{0}$ is the least element and members of D are ordered by $w_1 \sqsubseteq w_2$ iff $w_1 \oplus w_2 = w_2$.)

A *weighted SPKI/SDSI system* $\mathcal{WSS}$ is a 3-tuple $(\mathcal{C}, \mathcal{S}, f)$, where $\mathcal{C}$ is a set of certs, $\mathcal{S} = (D, \oplus, \otimes, \bar{0}, \bar{1})$ is a bounded idempotent semiring, and $f : \mathcal{C} \to D$ assigns weights to the certs in $\mathcal{C}$. We extend the function f to certificate chains in a natural way, i.e., given a certificate chain $\langle c_1, c_2, \cdots, c_k \rangle$, $f(\langle c_1, c_2, \cdots, c_k \rangle)$ is defined as $f(c_1) \otimes f(c_2) \otimes \cdots \otimes f(c_k)$.

Definition 4.3.2 *Given a weighted SPKI/SDSI system* $\mathcal{WSS} = (\mathcal{C}, \mathcal{S}, f)$ *and a request* $r = (K_U, R, T_R)$, chains$(\mathcal{C}, r)$ *denotes the set of certificate chains that are candidates for proving that request* r *can be fulfilled. Formally,* chains$(\mathcal{C}, r)$ *is the set of certificate chains* $\langle c_1, c_2, \cdots, c_k \rangle$ *that do not contain any invalid certificates, such that*

$$(c_k \circ \cdots \circ c_2 \circ c_1)(K_{\text{owner}[R]} \;\square) \in \{K_U \;\square, K_U \;\blacksquare\}.$$

The generalized authorization problem (GAP) is to answer the following questions and determine the indicated quantities:

(1) *Is* chains$(\mathcal{C}, r)$ *non-empty?*

(2) *If* chains$(\mathcal{C}, r)$ *is non-empty,*

- *Find the value* $\delta = \bigoplus \{ f(cc) \mid cc \in \text{chains}(\mathcal{C}, r) \}$.
- *Test whether* $T_R \sqsubseteq \delta$, *and if so, find a finite* witness set *of certificate chains* $\omega \subseteq \text{chains}(\mathcal{C}, r)$ *such that* $\bigoplus_{cc \in \omega} f(cc) = \delta$.

The GAP is satisfiable *if* $T_R \sqsubseteq \delta$, *in which case* ω *represents a proof of the problem's satisfiability.* $\square$

Solving a GAP requires finding a set of certificate chains such that the combine of the weights (using the operator $\oplus$) is maximal. Note that the extend operation $\otimes$ is used to calculate the value of a certificate chain.[7] The value of a set of certificate chains is computed using the combine operation $\oplus$. In general, it is enough for ω to contain only a finite set of maximal elements (i.e., maximal with respect to the semiring's partial order, $w_1 \sqsubseteq w_2$ iff $w_1 \oplus w_2 = w_2$).

The general strategy for solving a GAP algorithmically is as follows: the set of labeled SPKI/SDSI certificates is first translated to a weighted pushdown system.[8] After the translation, the answer is obtained by solving a generalized reachability problem on the WPDS, using the techniques presented in §2.2, to determine whether a set of certificate chains exists that has an appropriate (combined) weight.

---

[7]In a bounded idempotent semiring, $\otimes$ is not required to be commutative. The ability to use noncommutative $\otimes$ operators is important in the program-analysis applications discussed in §2.2; however, in trust-management applications, all of the different $\otimes$ operators have so far turned out to be commutative (cf. Tab. 4.1).

[8]In a GAP, each certificate is labeled with a weight. However, a label might depend on some global property. For example, for recency policies, a certificate's value represents the time the certificate was issued or last known to be current.

	Certificates	Weights	WPDS Rules
(1)	$K_X \; \square \; \longrightarrow \; K_H \text{ patient } \blacksquare$	I	$\langle K_X, \square \rangle \xrightarrow{I} \langle K_H, \text{patient } \blacksquare \rangle$
(2)	$K_H \text{ patient } \longrightarrow \; K_{H-AIDS} \text{ patient}$	I	$\langle K_H, \text{patient} \rangle \xrightarrow{I} \langle K_{H-AIDS}, \text{patient} \rangle$
(3)	$K_H \text{ patient } \longrightarrow \; K_{H-IM} \text{ patient}$	I	$\langle K_H, \text{patient} \rangle \xrightarrow{I} \langle K_{H-IM}, \text{patient} \rangle$
(4)	$K_{H-AIDS} \text{ patient } \longrightarrow \; K_{Charles}$	S	$\langle K_{H-AIDS}, \text{patient} \rangle \xrightarrow{S} \langle K_{Charles}, \varepsilon \rangle$
(5)	$K_{H-IM} \text{ patient } \longrightarrow \; K_{Charles}$	I	$\langle K_{H-IM}, \text{patient} \rangle \xrightarrow{I} \langle K_{Charles}, \varepsilon \rangle$

Figure 4.5: A set of weighted certificates and their corresponding weighted PDS rules.

Example 4.3.3 Privacy-preserving certificate chains. We now return to the example described in §4.3.1, in which company X offers a special insurance product to patients of a certain hospital H. The certificates relevant to the problem are shown in Fig. 4.5. $K_X \square$ represents the product offered, i.e., the additional insurance offered by company X. The filled square represents the fact that this authorization cannot be delegated, e.g., an eligible patient cannot delegate the permission to buy insurance to one of their friends. The principals corresponding to the AIDS and internal-medicine clinics in hospital H are denoted by K_{H-AIDS} and K_{H-IM}. Charles is a patient in both clinics.

Suppose that Charles wants to buy the insurance. In this case, $(4) \circ (2) \circ (1)$ equals $K_X \square \xrightarrow{S} K_{Charles} \blacksquare$, and $(5) \circ (3) \circ (1)$ equals $K_X \square \xrightarrow{I} K_{Charles} \blacksquare$; hence, either of the certificate chains $\langle (1), (2), (4) \rangle$ or $\langle (1), (3), (5) \rangle$ is sufficient to demonstrate that Charles is authorized to purchase the insurance product offered. However, the certificate chain $\langle (1), (2), (4) \rangle$ reveals that Charles is a patient in the AIDS clinic, which is information that Charles may not wish to reveal to company X. Therefore, Charles would prefer to offer the certificate chain $\langle (1), (3), (5) \rangle$ to company X; it proves that he is authorized to buy the additional insurance, but reveals the least amount of information about him.

Because we are concerned with preserving privacy, the weights associated with the rewrite rules will be a measure of the amount of information that they reveal. Because a GAP solution represents a $\oplus$ (join) over the weights of all certificate chains, rule (4)—which reveals the sensitive information that Charles receives treatment for AIDS—will carry a smaller weight (in the sense of the semiring partial order $\sqsubseteq$) than rule (5).[9] The problem we then want to solve is to find a set of certificate chains from $\langle K_H, \square \rangle$ to $\langle K_{Charles}, \blacksquare \rangle$ for which the combine ($\oplus$) of the weights of the chains is maximal. This biases certificate-chain discovery away from using certificates whose weight labels them as revealing sensitive information.

Privacy can be modeled in the GAP framework using the semiring $(D, \oplus, \otimes, \overline{0}, \overline{1})$, defined as follows: $D = \{Z, S, I\}$, where Z is the $\overline{0}$ element, and S and I stand for "sensitive" and "insensitive", respectively. The $\overline{1}$ element is S. The $\oplus$ and $\otimes$ operators are defined as follows:

$\oplus$	Z	S	I
Z	Z	S	I
S	S	S	I
I	I	I	I

$\otimes$	Z	S	I
Z	Z	Z	Z
S	Z	S	S
I	Z	S	I

It is easy to check that conditions 1–4 of Defn. 2.2.6 are satisfied. Condition 5 is trivially satisfied because D is finite.

The corresponding pushdown system $\mathcal{P}_C$ has the control locations $\{K_X, K_H, K_{H-AIDS}, K_{H-IM}, K_{Charles}\}$, the stack alphabet $\{\texttt{patient}, \square, \blacksquare\}$, and the set of rules listed in column 4 of Fig. 4.5. Rule (4), $K_{H-AIDS} \texttt{ patient} \xrightarrow{S} K_{Charles}$, is labeled S because it reveals that Charles is a patient in the AIDS clinic; all of the other certificates are labeled

[9]Formally, the weight domain would have three possible values, which are totally ordered as follows: $Z \sqsubset S \sqsubset I$; thus Z is the $\overline{0}$ element of the semiring.

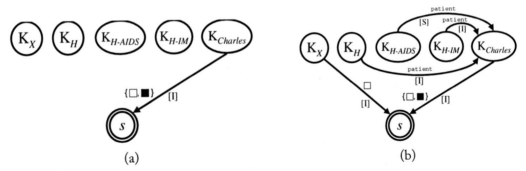

Figure 4.6: (a) Automaton that represents the configurations $Q = \{\langle K_{Charles}, \square \rangle, \langle K_{Charles}, \blacksquare \rangle\}$. (b) Automaton that represents the configurations in $pre^*(Q)$.

	D	⊕	⊗	$\overline{0}$	$sr\,One$
Privacy	$\{Z, S, I\}$	⊔	⊓	Z	I
Resources	$\mathcal{P}(\text{Resources})$	∪	∩	$\emptyset$	Resources
Validity	$\mathbb{N} \cup \{\pm\infty\}$	max	min	$-\infty$	$+\infty$
Recency	$\mathbb{N} \cup \{\infty\}$	min	max	∞	0
Trust	$\{Z, L, M, H\}$	⊔	⊓	Z	H

I. The weights of the certificate chains $\langle (1), (2), (4) \rangle$ and $\langle (1), (3), (5) \rangle$ are $I \otimes I \otimes S = S$ and $I \otimes I \otimes I = I$, respectively. Obviously, Charles prefers the certificate chain with weight I.

By using the techniques for generalized pushdown reachability on WPDSs described in §2.2 [239], it is possible for an authorization system to discover such certificate chains and their weights. One way to do this is to solve the generalized prestar query $pre^*(Q)$, where $Q = \{\langle K_{Charles}, \square \rangle, \langle K_{Charles}, \blacksquare \rangle\}$. The automaton shown in Fig. 4.6(a) accepts the set Q. The set $pre^*(Q)$ is shown in Fig. 4.6(b). Because there is a transition labeled I on the symbol $\square$ from state K_X to the accepting state s, $\delta(\langle K_X, \square \rangle) = I$. $\square$

4.3.5 USING SEMIRINGS IN OTHER KINDS OF AUTHORIZATION SPECIFICATIONS

We now discuss how several other authorization-related problems can be cast in the generalized authorization framework. Semirings for various authorization problems are shown in Tab. 4.1. These semiring labels can help us answer several useful questions, such as, "How long does a specific authorization last?" and "What is the trust level associated with an authorization?".

4.3.5.1 Maximally-Valid Certificate Chains

Let $V(c)$ be the expiration value of cert c, i.e., the cert c will expire at time $T_{current} + V(c)$, where $T_{current}$ is the current time. The expiration value of a certificate chain $\langle c_1\ c_2\ \cdots\ c_k \rangle$ is $\min_{i=1}^{k} V(c_i)$. Suppose that Alice wants to login to host H. If Alice provides a certificate chain that is only valid for two minutes, then she will be logged off by the host after two minutes. Thus, Alice wants to find a certificate chain that not only authorizes her to login to H but has the maximum expiration value among all such certificate chains. This is captured by the semiring $(\mathbb{N} \cup \{\pm\infty\}, \max, \min, -\infty, +\infty)$.[10]

We label each cert with the interval representing its validity period and run Algorithm *prestar*. Let i be the label associated with the transition $(K_{\text{owner}[H]}, \Box, s)$ in the automaton produced by Algorithm *prestar*. Then the authorization for Alice to login to host H will be valid for i time units. (The reference monitor for H can use this information to log Alice off after i time units.)

4.3.5.2 Most-Recent Certificate Chains

Let $R(c)$ be the time (relative to the current time) when cert c was issued, or an on-line check was performed on cert c; i.e., $T_{current} - R(c)$ is the actual time of issue or the last on-line check. We call $R(c)$ the *recency* associated with cert c. The recency of a certificate chain $\langle c_1, c_2, \cdots, c_k \rangle$ is equal to $\max_{i=1}^{k} R(c_i)$. Suppose that Alice wants to login to host H. For risk-reduction purposes, host H might mandate the use of a certificate chain whose recency is no more than ten minutes. In this case, Alice wishes to find a certificate chain that authorizes her to login to H and has the minimum recency among all such chains. Let $\langle c_1, c_2, \cdots, c_k \rangle$ be the certificate chain with minimum recency. If $\max_{i=1}^{k} R(c_i)$ is less than or equal to ten minutes, then Alice can use the certificate chain to login to H.

4.3.5.3 Certificate Chains with Maximal Trust Levels

For this problem, each certificate c is assigned a trust level $\mathrm{Tr}(c)$ by the issuer of the certificate. Intuitively, $\mathrm{Tr}(c)$ denotes the confidence that the issuer of c has in the relationship expressed by certificate c. The trust level of a certificate chain $\langle c_1, c_2, \cdots, c_k \rangle$ is $\bigotimes_{i=1}^{k} Tr(c_i)$,

Suppose that Alice wants to use server S, but S requires a certificate chain that is more trusted than a certain value v. In this case, Alice wants to find a certificate chain that is the most trusted among all potential certificate chains. If such a certificate chain has a trust value w that is more trusted than v (i.e., $v \sqsubseteq w$ in the semiring ordering), the reference monitor will let Alice use S.

The elements of the semiring represent trust levels, where low, medium, and high trust levels are denoted by $L, M,$ and H, respectively. A fourth value, Z, is added as the $\overline{0}$ element. $Z, L, M,$ and H form a totally ordered set, where $Z \sqsubset L \sqsubset M \sqsubset H$. The semiring operations $\oplus$ and $\otimes$ correspond

---

[10]$(\mathbb{N} \cup \{\pm\infty\}, \max, \min, -\infty, +\infty)$ has infinite ascending chains; however, the only operations performed are min and max. Hence, only a finite number of values ever arise in any run of the saturation process, and the semiring-labeling framework still applies. In general, it is acceptable to have infinite ascending chains as long only a finite number of values ever arise in any run of the saturation process [239]. For instance, regardless of whether we think of the set of resources as being of finite or infinite cardinality, the authorization specifications of a given set of certificates can only name a finite number of sets, durations of valid certificates, and roots of tree-structured file hierarchies, and so only a finite number of values ever arise during saturation.

to join ($\sqcup$) and meet ($\sqcap$) on the totally ordered set $\{Z, L, M, H\}$:

$$x \sqcup y = \begin{cases} x & \text{if } x \sqsupseteq y \\ y & \text{otherwise} \end{cases} \qquad x \sqcap y = \begin{cases} y & \text{if } x \sqsupseteq y \\ x & \text{otherwise} \end{cases}$$

When a trust level drawn from $\{Z, L, M, H\}$ is assigned to each certificate, and $\otimes$ is $\sqcap$, the trust level of a certificate chain $\langle c_1, c_2, \cdots, c_k \rangle$, namely $\bigotimes_{i=1}^{k} Tr(c_i)$, represents the trust level of the least-trusted link in the certificate chain. (Note that, due to the orientation of the semiring—i.e., $Z \sqsubset L \sqsubset M \sqsubset H$—"least-trusted link" in the sense of everyday speech corresponds to "link with the smallest semiring value".) For example, if the trust level of a chain is H, then all of the chain's labels must have the value H.

Because H, which denotes "highest level of trust", is highest in the total order, certificate-chain discovery is biased away from using certs with low levels of trust.

4.3.6   DISCUSSION

4.3.6.1 SPKI/SDSI and WPDSs

§2.2 discussed analysis techniques for overapproximating the set of reachable states in a program, and talked about its potential use for identifying security vulnerabilities in programs. §4.3.4 discussed the generalized authorization problem for SPKI/SDSI. While these two topics might, at first blush, seem unrelated, at the technical level, they turn out to be closely related: problems in both areas can be formulated using the same machinery, namely, weighted pushdown systems. WPDS solvers represent a united technology for the key algorithms required in both areas.

To some extent, these ideas were developed together, which had an unexpected payoff because improvements motivated by the needs in one area had serendipitous benefits in the other area:

Program Analysis → Trust Management. Pre*, post*, and LTL model-checking algorithms for (ordinary unweighted) PDSs were originally developed for analysis of programs. In particular, these algorithms provide a way to obtain finite representations of certain infinite sets of configurations.

However, a similar kind of issue arises in SPKI/SDSI: Given a set of certificates $\mathcal{C}$, its *closure* (denoted by $\mathcal{C}^\star$) is the smallest set of certificates that includes $\mathcal{C}$ and is closed under composition. In general, however, $\mathcal{C}^\star$ is infinite and hence cannot be represented explicitly. For example, consider the set of certificates $\mathcal{C} = \{(K \text{ A} \longrightarrow K \text{ A A})\}$. The closure $\mathcal{C}^\star$ of $\mathcal{C}$ is the following set:

$$\{(K \text{ A} \longrightarrow K \text{ A}^i) : i \geq 2\}$$

Given a name N and a set of certificates $\mathcal{C}$, $V_{\mathcal{C}}(N)$ is defined as

$$V_{\mathcal{C}}(N) = \mathcal{C}^\star(N) \cap \mathcal{K}.$$

In other words, $V_{\mathcal{C}}(N)$ is the set of keys that can be obtained from N by using the rewrite rules corresponding to the set of certs $\mathcal{C}$. In applications, if N is granted a certain authorization, (roughly) every key in $V_{\mathcal{C}}(N)$ is also indirectly granted that authorization.

The existing PDS algorithms allowed Jha and Reps to address many SPKI/SDSI problems using PDSs [141; 142]. In particular, they gave a name-resolution algorithm for SPKI/SDSI that improved on a special-purpose algorithm that had been developed previously [69], and also furnish algorithms for a large number of certificate-set vulnerability-analysis problems that had never been previously addressed.

Trust Management → Program Analysis. Further consideration of the needs that arise in trust-management systems (such as privacy, authorization expiration, accounting for recency of certificates, and finding a maximal-trust authorization) led Schwoon et al. [239] to introduce weighted PDSs, which then allowed them to devise new algorithms for interprocedural dataflow analysis that answered a much broader class of queries—such as stack-qualified queries—than any previously known algorithm [217].

Program Analysis → Trust Management. The first cut at formulating a way to extend PDSs with weights considered only weight domains that were totally ordered. The fact that program-analysis problems require weight domains with partial orders led us to consider a more general class of weighted pushdown systems [217; 239].[11] Partially ordered weight domains then turned out to have applications in certificate-set analysis [142, Sect. 4.6] and, in particular, resolve the issue raised by Li and Mitchell [159] that the name-resolution algorithm of Clarke et al. [69] is incomplete. Algorithms to solve path problems in WPDSs provide complete name-resolution algorithms for SPKI/SDSI (and run in polynomial time).

4.3.6.2 Incompleteness of Earlier Methods for Certificate-Chain Discovery

Clarke et al. [69] introduced the following heuristic to remove all "useless" certificates before carrying out certificate-chain discovery:

Heuristic 4.3.4 [69]

(a) Remove every name and auth cert that has an invalid validity specification (e.g., an expired validity specification).

(b) Remove every auth cert $C = (K, S, D, T, V)$ for which T is not equal to or greater than the authorization specification T' of the request (i.e., for which $T \not\supseteq T'$).

□

The elimination of useless certs by Heuristic 4.3.4(b) ensures that a chain found during certificate-chain discovery justifies granting the authorization T'. However, the approach has a significant drawback, which is that it does not handle situations in which a proof of authorization requires multiple certificate chains, each of which proves some *part* of the required authorization).[12]

---

[11]The special case of totally ordered weights turns out to be solvable more efficiently than the partially ordered case. Hence, our original approach "survived" and is discussed in [239, Sect. 6.5].

[12]This is the basis for the observation by Li and Mitchell that the "5-tuple reduction rule" of [106] is incomplete [159]. Although [69] speaks of "certificate chains", and [106] refers to "... finding the correct list of reductions ...", which appears to imply a single path; Carl Ellison confirmed to us that a scenario that requires justification via a set of certificate chains (or a dag of certificates) is reasonable, and noted that the justification of k-of-m threshold subjects also requires a set or a dag [105].

Example 4.3.5 For instance, consider the following certificate set:

$$c_1 : (K \ K_A \ 0 \ ((\text{dir /afs/cs.wisc.edu/public/tmp}) \ \text{read}) \ [t_1 \ldots t_2])$$
$$c_2 : (K \ K_A \ 0 \ ((\text{dir /afs/cs.wisc.edu/public/tmp}) \ \text{write}) \ [t_1 \ldots t_2])$$

Suppose that Alice makes the request

$$(K_A, ((\text{dir /afs/cs.wisc.edu/public/tmp}) \ (* \ \text{set read write}))).$$

In this case, the chain $\langle c_1 \rangle$ authorizes Alice to read from directory /afs/cs.wisc.edu/public/tmp, and a separate chain $\langle c_2 \rangle$ authorizes her to write to /afs/cs.wisc.edu/public/tmp. Together, $\langle c_1 \rangle$ and $\langle c_2 \rangle$ prove that she has both read and write privileges for /afs/cs.wisc.edu/public/tmp. However, both of the certificates c_1 and c_2 would be removed from the certificate set prior to running the certificate-chain discovery algorithm because $\{\text{read}\} \not\supseteq \{\text{read}, \text{write}\}$ and $\{\text{write}\} \not\supseteq \{\text{read}, \text{write}\}$. Consequently, no proof of authorization for Alice's request would be found. □

The algorithm for solving GAPs is more powerful than the certificate-chain-discovery methods developed by Clarke et al. [69] and Jha and Reps [142]. The problem defined in Defn. 4.3.2 asks for a witness *set* of certificate chains to be identified, which addresses the issue raised in Ex. 4.3.5. This is what allows the approach discussed in §4.3.4 to dispense with the elimination of useless certs by Heuristic 4.3.4(b); in particular, the $\oplus$ operation applied to the weights from multiple chains in witness set ω—for which each chain cc *individually* does not satisfy $f(cc) \sqsupseteq T'$—can yield a weight $\bigoplus_{cc \in \omega} f(cc) = \delta \sqsupseteq T'$ that satisfies the GAP.

For instance, in the case discussed in Ex. 4.3.5, we would work with a semiring in which the weights are drawn from the powerset of access-privileges, and the semiring ordering relation $\sqsubseteq$ is $\subseteq$. The GAP is satisfiable because the witness set $\{\langle c_1 \rangle, \langle c_2 \rangle\}$ discussed in Ex. 4.3.5 has the weight

$$f(\langle c_1 \rangle) \oplus f(\langle c_2 \rangle) = \{\text{read}\} \cup \{\text{write}\}$$
$$\supseteq \{\text{read}, \text{write}\}.$$

4.4   RT

The RT family of Role-based Trust-management languages were introduced in [161]. More specifically, we consider RT[←, ∩] and its three sub-languages: RT[ ], RT[←], and RT[∩].

SYNTAX

The basic constructs of RT[←, ∩] are *principals* and *role names*. We use A, B, D, E, F, X, Y, and Z, sometimes with subscripts, to denote principals. A role name is a word over some given standard alphabet. We use r, u, and w, sometimes with subscripts, to denote role names. A *role* takes the form of a principal followed by a role name, separated by a dot, e.g., $A.r$ and $X.u$. A *role* defines a set of principals that are members of this role. Each principal A has the authority to designate the members of each role of the form $A.r$. An access control permission is represented as a role as well; for example, making B a member of $A.r$ may mean that B has permission to do action r on the object A.

There are four types of policy statements in $RT[\leftarrow, \cap]$, each corresponding to a different way of defining role membership. Each statement has the form $A.r \longleftarrow e$, where $A.r$ is a role and e is a role expression, to be defined below. We read "$\longleftarrow$" as "includes", and say the policy statement *defines* the role $A.r$.

1. *Simple Member*: $A.r \longleftarrow D$

   This statement means that A asserts that D is a member of A's r role.

2. *Simple Inclusion*: $A.r \longleftarrow B.r_1$

   This statement means that A asserts that its r role includes (all members of) B's r_1 role. This represents a delegation from A to B, as B may add principals to become members of the role $A.r$ by issuing statements defining $B.r_1$.

3. *Linking Inclusion*: $A.r \longleftarrow A.r_1.r_2$

   We call $A.r_1.r_2$ a *linked role*. This statement means that A asserts that $A.r$ includes $B.r_2$ for every B that is a member of $A.r_1$. This represents a delegation from A to all the members of the role $A.r_1$.

4. *Intersection Inclusion*: $A.r \longleftarrow B_1.r_1 \cap B_2.r_2$

   We call $B_1.r_1 \cap B_2.r_2$ an *intersection*. This statement means that A asserts that $A.r$ includes every principal who is a member of both $B_1.r_1$ and $B_2.r_2$. This represents partial delegations from A to B_1 and to B_2.

A *role expression* is a principal, a role, a linked role, or an intersection. Given a set $\mathcal{P}$ of policy statements, we define the following: $Principals(\langle\rangle)$ is the set of principals in $\mathcal{P}$, $Names(\langle\rangle)$ is the set of role names in $\mathcal{P}$, and $Roles(\langle\rangle) = \{A.r \mid A \in Principals(\langle\rangle), r \in Names(\langle\rangle)\}$. $RT[\leftarrow, \cap]$ is a slightly simplified (yet expressively equivalent) version of RT_0 [164].

We consider also the following sub-languages of $RT[\leftarrow, \cap]$: $RT[\,]$ has only simple member and simple inclusion statements, $RT[\leftarrow]$ adds to $RT[\,]$ linking inclusion statements, and $RT[\cap]$ adds to $RT[\,]$ intersection inclusion statements.

Example 4.9   Consider the following scenario. The system administrator of a company, SA, controls access to some resource, which we abstractly denote by SA.access. The company policy is the following: managers always have access to the resource, managers can delegate the access to other principals but only to employees of the company, and HR is trusted for defining employees and managers.

The state $\mathcal{P}$ consists of the following statements:

$$\text{SA.access} \longleftarrow \text{SA.manager} \tag{1}$$
$$\text{SA.access} \longleftarrow \text{SA.delegatedAccess} \cap \text{HR.employee} \tag{2}$$
$$\text{SA.manager} \longleftarrow \text{HR.manager} \tag{3}$$
$$\text{SA.delegatedAccess} \longleftarrow \text{SA.manager.access} \tag{4}$$
$$\text{HR.employee} \longleftarrow \text{HR.manager} \tag{5}$$
$$\text{HR.employee} \longleftarrow \text{HR.programmer} \tag{6}$$
$$\text{HR.manager} \longleftarrow \text{Alice} \tag{7}$$
$$\text{HR.programmer} \longleftarrow \text{Bob} \tag{8}$$
$$\text{HR.programmer} \longleftarrow \text{Carl} \tag{9}$$
$$\text{Alice.access} \longleftarrow \text{Bob} \tag{10}$$

Given the state $\mathcal{P}$ above, we have.

$$
\begin{aligned}
Principals(\langle\rangle) &= \{\text{SA, HR, Alice, Bob, Carl}\} \\
Names(\langle\rangle) &= \{\text{access, manager, delegatedAccess, employee, programmer}\} \\
Roles(\langle\rangle) &= \{A.r \mid A \in Principals(\langle\rangle), r \in Names(\langle\rangle)\} \\
&= \{\text{SA.access, SA.manager,} \cdots, \text{Carl.programmer}\}
\end{aligned}
$$

SEMANTICS

We give a formal characterization of the meaning of a set $\mathcal{P}$ of policy statements by translating each policy statement into a datalog clause. We call the resulting program the *semantic program* of $\mathcal{P}$.

Definition 4.10 Semantic Program. Given a set $\mathcal{P}$ of policy statements, the *semantic program*, $SP(\mathcal{P})$, of $\mathcal{P}$, has one ternary predicate m. Intuitively, $m(A, r, D)$ means that D is a member of the role $A.r$. $SP(\mathcal{P})$ is the set of all datalog clauses produced from policy statements in $\mathcal{P}$. The rules to generate the Semantic Program $SP(\mathcal{P})$ from $\mathcal{P}$ are shown below. Symbols that start with "?" represent logical variables.

For each $A.r \longleftarrow D$ in $\mathcal{P}$, add
$$m(A, r, D) \tag{m1}$$
For each $A.r \longleftarrow B.r_1$ in $\mathcal{P}$, add
$$m(A, r, ?Z) :- m(B, r_1, ?Z) \tag{m2}$$
For each $A.r \longleftarrow A.r_1.r_2$ in $\mathcal{P}$, add
$$m(A, r, ?Z) :- m(A, r_1, ?Y), m(?Y, r_2, ?Z) \tag{m3}$$
For each $A.r \longleftarrow B_1.r_1 \cap B_2.r_2$ in $\mathcal{P}$, add
$$m(A, r, ?Z) :- m(B_1, r_1, ?Z), m(B_2, r_2, ?Z) \tag{m4}$$

We write $SP(\mathcal{P}) \models m(X, u, Z)$ when $m(X, u, Z)$ is in the minimal Herbrand model of $SP(\mathcal{P})$.

QUERIES

We consider the following three forms of query Q:

1. *Membership*: $A.r \sqsupseteq \{D_1, \ldots, D_n\}$

   Intuitively, this means that all the principals $D_1, \ldots, D_n$ are members of $A.r$. Formally, $\mathcal{P} \vdash A.r \sqsupseteq \{D_1, \ldots, D_n\}$ if and only if $\{Z \mid SP(\mathcal{P}) \models m(A, r, Z)\} \supseteq \{D_1, \ldots, D_n\}$.

2. *Boundedness*: $\{D_1, \ldots, D_n\} \sqsupseteq A.r$

   Intuitively, this means that the member set of $A.r$ is bounded by the given set of principals. Formally, $\mathcal{P} \vdash \{D_1, \ldots, D_n\} \sqsupseteq A.r$ if and only if $\{D_1, \ldots, D_n\} \supseteq \{Z \mid SP(\mathcal{P}) \models m(A, r, Z)\}$.

3. *Inclusion*: $X.u \sqsupseteq A.r$

   Intuitively, this means that all the members of $A.r$ are also members of $X.u$. Formally, $\mathcal{P} \vdash X.u \sqsupseteq A.r$ if and only if $\{Z \mid SP(\mathcal{P}) \models m(X, u, Z)\} \supseteq \{Z \mid SP(\mathcal{P}) \models m(A, r, Z)\}$.

Example 4.11   If $\mathcal{P}$ is the state given in Example 4.9, the following queries yield the indicated results:

Membership:	$\mathcal{P} \vdash$ SA.access $\sqsupseteq$ {Eve}	(False)
Membership:	$\mathcal{P} \vdash$ SA.access $\sqsupseteq$ {Alice}	(True)
Boundedness:	$\mathcal{P} \vdash$ {Alice, Bob} $\sqsupseteq$ SA.access	(True)
Inclusion:	$\mathcal{P} \vdash$ HR.employee $\sqsupseteq$ SA.access	(True)

RESTRICTION RULES ON STATE CHANGES

Using statements in RT[$\leftarrow, \cap$], one can delegate control over resources to other principals. In Example 4.9, the two statements SA.access$\longleftarrow$SA.delegatedAccess $\cap$ HR.employee and SA.delegatedAccess $\longleftarrow$ SA.manager.access together mean that any principal that is a manager can affect who can access the resource. For example, Alice could add Alice.access $\longleftarrow$ Carl giving Carl access. In the resulting state $\mathcal{P}'$, $\mathcal{P}' \vdash$ {Alice, Bob} $\sqsupseteq$ SA.access is false, whereas the result is true for $\mathcal{P}$. From the System Administrator (SA)'s perspective, roles such as Alice.access are not under its control. New statements defining Alice.access may be issued by Alice and existing statements defining Alice.access may be revoked. In order for SA to understand the effect of the two statements mentioned above, SA may want to know whether some desirable security properties always hold even though statements defining roles such as Alice.access can be changed arbitrarily.

We now present a concrete formulation of restriction rules that enable one to articulate analysis questions concerning the states that are reachable based on changes to policy. To model control over roles, we use restriction rules of the form $\mathcal{R} = (\mathcal{G}_\mathcal{R}, \mathcal{S}_\mathcal{R})$, which consist of a pair of finite sets of roles. (In the rest of the paper, we drop the subscripts from $\mathcal{G}$ and $\mathcal{S}$, as $\mathcal{R}$ is clear from context.)

1. Roles in $\mathcal{G}$ are called *growth-restricted* (or *g-restricted*); no policy statements defining these roles can be added. Roles not in $\mathcal{G}$ are called *growth-unrestricted* (or *g-unrestricted*).

2. Roles in $\mathcal{S}$ are called *shrink-restricted* (or *s-restricted*); policy statements defining these roles cannot be removed. Roles not in $\mathcal{S}$ are called *shrink-unrestricted* (or *s-unrestricted*).

We would like to point out that, if a role $A.r$ that is g-restricted is defined to include a role $B.r_1$ that is g-unrestricted, then no new statement defining $A.r$ can be added; however, new statements defining $B.r_1$ can be added, indirectly adding new members to $A.r$. This effect makes security analysis more challenging.

An example of $\mathcal{R}$ is $(\emptyset, Roles(\langle\rangle))$, under which every role may grow without restriction, and no statement defining roles in $Roles(\langle\rangle)$ can be removed. This models the case of having incomplete knowledge of a global policy state. In this case, one sees a set $\mathcal{P}$ of statements but thinks that there are other statements in the global state that are currently unknown, and one wants to know whether certain security properties always hold no matter what these unknown statements may be.

Another example is $\mathcal{R} = (\mathcal{G}, \mathcal{S})$, where $\mathcal{G} = \mathcal{S} = \{X.u \mid X \in \{X_1, \ldots, X_k\}, u \in Names(\langle\rangle)\}$. This corresponds to the scenario in which there are principals that are trusted and one wants to analyze the effect of policy changes of untrusted principals. Here $X_1, \ldots, X_k$ are identified as trusted, and other principals are not trusted.

If a principal X does not appear in the restriction rule $\mathcal{R}$, then for every role name r, by definition $X.r$ is g/s-unrestricted. This models that the roles of unknown principals may be defined arbitrarily.

Example 4.12 Referring again to the example in Example 4.9, consider the restriction rule $\mathcal{R}$ given as follows:

$$\mathcal{G} = \{ \text{SA.access, SA.manager, SA.delegatedAccess, HR.employee} \}$$
$$\mathcal{S} = \{ \text{SA.access, SA.manager, SA.delegatedAccess, HR.employee, HR.manager} \}$$

In this restriction rule, SA and HR are assumed to be trusted; however, $\mathcal{G}$ allows statements to be added defining HR.manager and HR.programmer. Thus adding such statements cannot invalidate any security property obtained by using $\mathcal{R}$.

Given the above $\mathcal{R}$, statements (1) to (7) cannot be removed, statements (8) to (10) may be removed, new statements defining roles in $\mathcal{G}$ cannot be added, and one can add new statements defining HR.manager, HR.programmer, Alice.access, Bob.access, Carl.access, etc. We now list some example analysis problem instances, together with the answers:

Simply safety analysis:	Is "SA.access $\sqsupseteq$ {Eve}" possible?	(Yes)
Simple availability analysis:	Is "SA.access $\sqsupseteq$ {Alice}" necessary?	(Yes)
Bounded safety analysis:	Is "{Alice, Bob} $\sqsupseteq$ SA.access" necessary.	(No)
Containment analysis:	Is "HR.employee $\sqsupseteq$ SA.access" necessary?	(Yes)

Observe that the availability property "SA.access $\sqsupseteq$ {Alice} is necessary" depends on HR.manager being s-restricted. Together with our observations above concerning repeated analysis, this illustrates the advantage of allowing $\mathcal{G} \neq \mathcal{S}$.

USAGE OF SECURITY ANALYSIS

Security analysis can be used to help ensure that security requirements are met, and that they continue to be met after policy changes are made by autonomous, possibly malicious principals.

Let us say that a query $\mathcal{Q}$, a restriction rule $\mathcal{R}$, and a *sign*, either $+$ or $-$, together form a *requirement*. For instance, one requirement might consider whether everyone who has access to a particular confidential resource is an employee of the organization. In this case, the sign used would be $+$ to indicate that the condition should always hold, as this ensures that no one outside the organization has access to the confidential resource. A policy $\mathcal{P}$ *complies with* a requirement $\langle \mathcal{Q}, \mathcal{R}, + \rangle$ if $\mathcal{Q}$ is necessary given $\mathcal{R}$ and $\mathcal{P}$, and $\mathcal{P}$ complies with $\langle \mathcal{Q}, \mathcal{R}, - \rangle$ if $\mathcal{Q}$ is not possible given $\mathcal{R}$ and $\mathcal{P}$.

An organization's System Administrator (SA) writes a set of requirements based on a restriction rule $\mathcal{R}$ that forbids certain changes to certain roles. All roles g/s-restricted by $\mathcal{R}$ should be controlled by trusted principals in the organization. Assuming that we start in a policy state that complies with all the requirements, security analysis ensures that this compliance is preserved across changes to the policy state as long as these trusted principals cooperate as follows. When a change that is not allowed by $\mathcal{R}$ is made, it necessarily involves a trusted principal either adding a statement that defines a g-restricted role or removing a statement that defines a s-restricted role. When this occurs, the principal should first perform security analysis to determine whether the security requirements are met by the state that results from the prospective change, and then make the change only if that is the case. When a change that is allowed by $\mathcal{R}$ is made, be it by a trusted or an untrusted principal, then nothing special need be done, as such changes are taken into account by the analysis, and hence cannot lead to a state that is uncompliant with the requirements. When a change is made by a principal that is untrusted, the way $\mathcal{R}$ is chosen ensures that the change is made to a g/s-unrestricted role. Thus, the preservation of compliance does not depend on untrusted principals.

In this example, the SA determines a set of requirements based on a single restriction rule. In general, anyone may specify requirements it wishes to have maintained by the TM system. Different parties may have different sets of principals that they trust to assist in preserving the requirements, which will result in their using different restriction rules.

COMPLEXITY RESULTS

We now present a summary of the computational complexity results for security analysis in $RT[\leftarrow, \cap]$ and its sub-languages. Details of these results can be found in [162].

$RT[\leftarrow, \cap]$ and its sub-languages are monotonic, in the sense that adding more statements to a policy will always *increase* the set of implied membership facts. This important monotonicity property allows us to answer such analysis queries using datalog programs that run in polynomial time.

To answer a universal membership analysis instance that asks whether "$A.r \sqsupseteq \{D_1, \ldots, D_n\}$" is necessary given $\mathcal{P}$ and $\mathcal{R}$, one can consider the set of principals that are members of $A.r$ in every reachable state. We call this set the *lower-bound* of $A.r$. If the lower-bound of $A.r$ is a superset of $\{D_1, \ldots, D_n\}$, then the answer to the analysis is "yes"; otherwise, the answer is "no".

To compute the lower-bound of a role, consider the state obtained from $\mathcal{P}$ by removing all statements whose removal is permitted by $\mathcal{R}$. We denote this state by $\langle|\rangle_\psi$. Because $\mathcal{R}$ is static, the order of removing these statements does not matter, and $\langle|\rangle_\psi$ uniquely exists. Clearly, $\langle|\rangle_\psi$ is reachable; furthermore, $\langle|\rangle_\psi \subseteq \mathcal{P}'$ for every $\mathcal{P}'$ that is reachable from $\mathcal{P}$. As RT$[\leftarrow, \cap]$ is monotonic, the lower-bound of $A.r$ is the same as the set of principals who are members of the role $A.r$ in $\langle|\rangle_\psi$.

The lower-bound of $A.r$ can also be used to answer an existential boundedness analysis that asks whether "$\{D_1, \ldots, D_n\} \sqsupseteq A.r$" is possible given $\mathcal{P}$ and $\mathcal{R}$. If the lower-bound of $A.r$ is a subset of $\{D_1, \ldots, D_n\}$, then the answer is "yes"; otherwise, the answer is "no".

Existential membership (simple safety) analysis and universal boundedness (bounded safety) analysis can be answered by computing an "upper-bound" of role memberships. The upper-bound of a role is the set of principals that could become a member of the role in some reachable state. Intuitively, such bounds can be computed by considering a "maximal reachable state". However, such a "state" may contain an infinite set of policy statements, and the upper-bounds of roles may be infinite. Fortunately, one can simulate the upper bounds by a finite set and derive correct answers.

Inclusion queries are neither monotonic nor anti-monotonic. Given an inclusion query $X.u \sqsupseteq Z.w$ and three states $\mathcal{P}' \subseteq \mathcal{P} \subseteq \mathcal{P}''$, it is possible that $\mathcal{P} \vdash \mathcal{Q}$, but both $\mathcal{P}' \nvdash \mathcal{Q}$ and $\mathcal{P}'' \nvdash \mathcal{Q}$. Unlike membership and boundedness queries, we cannot simply look at a specific minimal (or maximal) state to answer an arbitrary inclusion query.

We restrict our attention to universal inclusion queries, as this is more interesting in terms of security properties than existential inclusion queries. We say that a role $X.u$ *contains* another role $A.r$ if $X.u \sqsupseteq A.r$ is necessary, i.e., $X.u$ includes $A.r$ in every reachable state. Observe that we use "contains" and "includes" as two technical terms that have different meanings. We call the problem of determining whether a role contains another role the *containment analysis* problem.

For RT[ ], which only allows membership and delegation policy statements, containment for all reachable states is computable by a stratified datalog program with negation in polynomial time. For RT[$\cap$], which is RT[ ] plus intersection, the problem becomes **coNP**-complete. Intuitively, the reason is that multiple statements about a role represent disjunction, while intersection of roles provides a corresponding form of conjunction. For RT[$\leftarrow$], which is RT[ ] plus role linking, role containment for all reachable policy states is **PSPACE**-complete. For RT[$\leftarrow, \cap$], which includes role linking, the problem remains decidable; it is shown in [162] that the problem is in **coNEXP** (or double-exponential time) and is **PSPACE**-hard.

CHAPTER 5

Analyzing Security Protocols

Protocols that enable secure communication over an untrusted network constitute an important part of the current computing infrastructure. Common examples of such protocols are SSL [111], TLS [94], Kerberos [151], and the IPSec [148] and IEEE 802.11i [8] protocol suites. SSL and TLS are used by internet browsers and web servers to allow secure transactions in applications like online banking. The IPSec protocol suite provides confidentiality and integrity at the IP layer and is widely used to secure corporate VPNs. IEEE 802.11i provides data protection and integrity in wireless local area networks, while Kerberos is used for network authentication.

The design and security analysis of such network protocols presents a difficult problem. In several instances, serious security vulnerabilities were uncovered in protocols many years after they were first published or deployed [61; 127; 169; 179; 181]. While some of these attacks rely on subtle properties of cryptographic primitives, a large fraction can be traced to intricacies in designing protocols that are robust in a concurrent execution setting. To further elaborate this point, let us consider the concrete example of the SSL protocol. In SSL, a client typically sets up a key with a web server. That key is then used to protect all data exchanged between them. A single client can simultaneously engage in sessions with multiple servers and a single server concurrently serves many clients. Let us consider a scenario in which all network traffic is under the control of the adversary. In addition, the adversary may also control some of the clients and servers. The protocol should guarantee certain security properties for honest agents even in such an adversarial environment. Specifically, if an honest client executes an SSL session with an honest server, the attacker should not be able to recover the exchanged key. This is called the *key secrecy* property. Furthermore, an attacker should not be able to fool an honest client into believing that she has completed a session with an honest server unless that is indeed the case. This property is called *authentication*. The security proof that SSL does indeed provide these guarantees, even when the cryptography is perfect, turns out to be far from trivial [128; 209]. The central problem is ensuring that the attacker cannot combine data acquired from a possibly unbounded number of concurrent sessions to subvert the protocol goals.

Over the last three decades, a variety of methods and tools have been developed for analyzing the security guarantees provided by network protocols. The main lines of work include specialized protocol logics [56; 88; 116; 247], process calculi [10; 11; 165; 212] and tools [178; 244], as well as theorem-proving [207; 208] and model-checking methods [169; 191; 224; 235] using general purpose tools. Our goal is to introduce our readers to this research area. This is, however, not a comprehensive survey of the field.

The chapter is structured as follows. The *general methodology of security protocol analysis* is described in Section 5.1. This section includes a discussion of the symbolic attacker model that is used in most protocol analysis methods. Section 5.2 describes a *specific protocol analysis method—Protocol Composition Logic (PCL)* [88]—that exemplifies the general methodology. The method is illustrated using a simple example along with citations to publications reporting on application to industrial protocols. Section 5.3 summarizes some of the other prominent methods for analysis of security protocols including model-checking, theorem-proving, process calculi, and symbolic reachability analysis, as well as decidability and complexity results about the protocol security problem. Section 5.4 reports on two significant problems that have been addressed in recent work. Progress on methods for *secure protocol composition* is reported in Section 5.4.1 while *cryptographically sound protocol analysis methods* are described in Section 5.4.2. Conclusions are presented in Section 5.5.

5.1 PROTOCOL ANALYSIS METHODOLOGY

In this section, we present a high-level overview of the general methodology for security protocol analysis. The next section presents a specific instance of this methodology. Protocol analysis consists of the following steps:

1. *Modeling the protocol and the adversary*:

   A *protocol* consists of a set of *roles*. For example, the SSL protocol consists of a *client* role and a *server* role. Each role can be viewed as a program, i.e. a sequence of actions such as sending and receiving messages, generating a new random number, performing cryptographic operations such as encryption and decryption, signature generation and verification and so on. In a typical deployment scenario, many instances of these roles are concurrently executed by different principals. For example, the server role of the SSL protocol is simultaneously executed by many web servers on the internet such as banking and e-commerce sites; the client role is executed simulatenously by many consumers who visit these websites. In addition to *honest principals* who are faithfully executing these programs, we have to allow for the possibility of an *adversary* who may act differently. The adversary can execute any number of role instances, e.g. it can execute SSL server programs by setting up its own web server or by compromising existing ones. Also, the adversary can intercept messages and inject messages that it can create into the network (e.g. by compromising routers). However, there are also some things that the adversary *cannot* do: she cannot break cryptography, e.g. she cannot decrypt an encrypted message if she does not possess the corresponding decryption key; and she cannot cause honest principals to deviate from faithfully following the protocol roles.

   A protocol analysis formalism has to capture all of the above mentioned aspects. There are 3 related questions:

   (a) How do we represent protocols?

   (b) How do we capture adversary capabilities?

   (c) How does a protocol execute in conjunction with an adversary?

One approach (exemplified in the next section for PCL) is to use a programming language. Protocols and adversaries are expressed as programs using the syntax of the language. Messages are represented using a term algebra. The execution of protocols and the capabilities of the adversary are captured by the operational semantics for the programming language. Process calculi (languages for expressing concurrent programs) have been used in a number of protocol analysis formalisms [10; 11; 165; 212]. Model-checking methods use finite state transition systems to represent both the protocol and the adversary; protocol execution follows the transition relation, which is often non-deterministic [169; 191; 224; 235]. Other approaches include inductive definitions for protocol executions (traces) used in theorem-proving work [207; 208] and term rewriting systems [102; 178; 223].

The adversary's capabilities include receiving all messages that are sent on the network and sending messages that it can compute using the following rules. The choice of which action to carry out is non-deterministic.

$$\Gamma \vdash m \wedge \Gamma \vdash k \implies \Gamma \vdash ENC_k\{\!|m|\!\}$$
$$\Gamma \vdash ENC_k\{\!|m|\!\} \wedge \Gamma \vdash \bar{k} \implies \Gamma \vdash m$$
$$\Gamma \vdash m \wedge \Gamma \vdash k \implies \Gamma \vdash SIG_k\{\!|m|\!\}$$
$$\Gamma \vdash m \implies \Gamma \vdash HASH\{\!|m|\!\}$$
$$\Gamma \vdash m_1 \wedge \Gamma \vdash m_2 \implies \Gamma \vdash m_1, m_2$$
$$\Gamma \vdash m_1, m_2 \implies \Gamma \vdash m_1 \wedge \Gamma \vdash m_2$$

Here $\Gamma \vdash m$ means that the adversary can compute message m from the set Γ of messages. These rules capture the idea that cryptography is perfect: An encrypted message can be decrypted only with the corresponding decryption key; signatures are unforgeable without access to the private signing key; hashes can be produced only if the adversary possesses the message and so on. In addition, the adversary can generate nonces and message constants. The adversary has complete control over the network: it can intercept every message sent on the network and send messages that it can construct (using the above inference rules) to honest parties. Finally, the adversary has an identity on the network, i.e. a name and corresponding private and public keys. This *symbolic adversary model* seems to have developed from positions taken by Needham-Schroeder [201], Dolev-Yao [97], and much subsequent work by others.

2. *Specifying security properties*:

Protocols are designed to provide security properties such as *authentication* and *secrecy*. Security properties capture certain desirable behaviors of the execution of a protocol in conjunction with an adversary. Their formal definitions thus necessarily depend on the execution model discussed in the previous item.

A widely used approach is to define what it means for a property to hold on a single execution or *trace*. A protocol is said to satisfy a property if it holds in *all* traces of the protocol. For example, secrecy of a message m may be defined to hold on a trace if the message is not sent out on the network in the clear in that trace; a protocol satisfies the secrecy property of a message if in all traces the message is not sent out in the clear. Similarly, authentication is often defined as a trace property that captures the idea that if agent A has completed a protocol trace supposedly with agent B, then B was indeed engaged in the session with A; this property has to again hold in all traces of the protocol.

A second approach is to specify properties using notions of *program equivalence*. Security properties are specified by requiring that the program for the protocol whose security we are trying to evaluate is equivalent to a program which is "obviously secure", e.g. because it uses private communication channels between principals to exchange secrets. This approach is adopted in a number of protocol analysis methods based on process calculi, e.g. spi-calculus [11], CryptoSPA [110], and the applied π−calculus [10]. Program equivalence can be defined in terms of trace equivalence, may and must testing, or by various forms of bisimulation. The interested readers are referred to the papers cited above for details.

3. *Checking for or proving property satisfaction*:

Given a formal model of a protocol and adversary (step 1), and a specification of the desired security properties (step 2), the final step in a protocol analysis method is to answer the following question: Does the protocol satisfy the desired properties? A variety of methods and tools have been developed to answer this question. The main lines of work include specialized protocol logics [56; 88; 116; 247], process calculi [10; 11; 165; 212] and tools [178; 244], as well as theorem-proving [207; 208] and model-checking methods [169; 191; 224; 235] using general purpose tools. There are several points of difference among these approaches. While most model-checking tools can only analyze a finite number of concurrent sessions of a protocol, some of the logics, process calculi, and theorem-proving techniques yield protocol security proofs without bounding the number of sessions; this is achieved using inductive and coinductive methods.

5.2 PROTOCOL COMPOSITION LOGIC

In this section, we present a specialized logic called *Protocol Composition Logic (PCL)* [87; 101; 221]. The presentation focuses on a fragment of the logic as described in a recent paper [88]. The formal system consists of (a) a language for modeling protocols and their execution in conjunction with an adversary, (b) a logic for specifying security properties, and (c) a proof system for proving that a given protocol satisfies a certain property. We provide below an overview of the three parts, as instances of the general methodology described in the previous section. The rest of the section provides additional formal details and uses PCL to carry out a proof of a signature-based challenge-response protocol.

1. *Modeling the protocol and the adversary*

A programming language is used to model protocols and their execution. This language is a conventional process calculus in the same vein as CCS, CSP, and their variants and descendants [133; 186]. The language consists of a *term algebra* for representing cryptographic messages, such as keys, nonces (random numbers), identities of principals, encryption, signatures; messages can also be concatenated. A *protocol role* (e.g. client role of SSL) is a sequence of *actions*, such as generating random numbers, sending and receiving messages, decryption and signature verification. A *protocol* is defined by a set of such roles. Multiple instances of protocol roles (i.e. *threads*) execute concurrently, together with an adversary. The execution of protocols is formalized using the operational semantics of the programming language and gives rise to *traces*, i.e. sequences of actions by different protocol participants and the adversary.

2. *Specifying security properties*

PCL formulas can express authentication and secrecy properties of protocols. The basic assertions are similar to Hoare logic [132] and dynamic logic [123], with the formula $\theta[P]_X\varphi$ stating that if actions P are executed in thread X, starting from a state where formula θ is true, then the formula φ is true about the resulting state. While the formula only mentions the actions P of thread X, states reached after X does P may arise as the result of these actions and any additional actions performed by other threads, including arbitrary actions by an attacker. PCL includes a number of action predicates, such as $\mathsf{Send}(X, t)$, $\mathsf{Receive}(X, t)$, $\mathsf{New}(X, t)$, $\mathsf{Decrypt}(X, t)$, $\mathsf{Verify}(X, t)$, which assert that the named thread has performed the indicated actions. The semantics of the logic is defined over traces of the protocol. For example, $\mathsf{Send}(X, t)$ holds on a trace if thread X sent the term t as a message. Authentication properties are specified by asserting that protocol participants sent and received messages in the order prescribed by the protocol, a form of *matching conversations* [38]. Secrecy properties are specified using the predicate $\mathsf{Has}(X, t)$, which intuitively means that t is built from constituents that X either generated (using a `new` action) or received in a way that did not hide them under encryption by a key not known by X. A protocol satisfies a security property if the property holds on all traces produced by executing the protocol.

3. *Checking for or proving property satisfaction*

The PCL proof system codifies and supports direct reasoning about the consequences of individual protocol steps, in a way that allows properties of individual steps to be combined to prove properties of complex protocols. The proof system is *sound*, i.e. if a property of a protocol is provable using the proof system, then it holds in all traces obtained by executing the protocol. The PCL axioms and inference rules fall into several categories. One class of axioms assert that after an action is performed, the indicated thread has performed that action. Another class of axioms state properties of cryptographic operations. For example, an axiom reflecting the unforgeability property of digital signatures states that whenever an agent verifies the signature of an honest agent, then that agent must have generated a signature on that message and sent it out in an earlier message.

PCL also uses a novel form of induction, currently referred to as the "honesty rule", in which induction over sequences of actions performed by honest agents can be used to derive conclusions about arbitrary runs in the presence of adversary actions. To see how this works, suppose that in some protocol, whenever a principal receives a message of the form $ENC_K\{a, b\}$, representing the encryption of a pair (a, b) under key K, the principal then responds with $ENC_K\{b\}$. Assume further that this is the only situation in which the protocol specifies that a message consisting of a single encrypted datum is sent. Using the honesty rule, it is possible to prove that if a principal A is honest, and A sends a message of the form $ENC_K\{b\}$, then A must have previously received a message of the form $ENC_K\{a, b\}$. For certain protocols, this form of reasoning allows us to prove that if one protocol participant completes the prescribed sequence of actions, and another principal named in one of the messages is honest, then the two participants are guaranteed a form of authentication.

Related Work.   The first generation of protocol logics, exemplified by BAN [56] and its successors—GNY and SvO logics [116; 248]—introduced several new ideas in the protocol analysis space. They allowed formal security proofs to be carried out at a high level of abstraction using a proof system—a collection of axioms and proof rules codifying basic facts about protocols. In contrast with model-checking, proofs in such logics did not involve explicit reasoning about attacker actions, guaranteed security for an unbounded number of concurrent protocol sessions and, in many cases, paralleled the high-level protocol design intuition. On the other hand, BAN logic also had some serious drawbacks. In particular, BAN is not sound with respect to the now standard symbolic model of protocol execution and attack, i.e. protocols that are proved secure using BAN logic could in fact be insecure. Another weakness of BAN logic is the need for an "abstraction phase" which involves representing a protocol using a set of logical formulas. Also, BAN uses non-standard logical concepts such as "jurisdiction" and supports proofs of authentication but not of secrecy. The interested reader is referred to the cited papers for further details on these logics.

PCL shares several features with the previous generation of protocol logics such as BAN and GNY [56; 116]: It was also initially designed as a logic of authentication; it annotates programs with assertions, does not require explicit reasoning about the actions of an attacker, and uses formulas for freshness, sending and receiving messages, and to express that two agents have a shared secret. In contrast to BAN and related logics, PCL avoids the need for an "abstraction" phase because PCL formulas contain the protocol programs, and PCL addresses temporal concepts directly, both through modal formulas that refer specifically to particular points in the execution of a protocol, and through temporal operators in pre- and post-conditions. PCL is also formulated using standard logical concepts (predicate logic and modal operators), does not involve "jurisdiction" or "belief", and has a direct connection with the execution semantics of network protocols that is used in explicit reasoning about actions of a protocol and an attacker, such as with Paulson's inductive method [208] and Schneider's rank function method [236]. Furthermore, it supports reasoning about secrecy.

$$A \rightarrow B \quad : \quad m$$
$$B \rightarrow A \quad : \quad n, SIG_B\{\!|n, m, A|\!\}$$
$$A \rightarrow B \quad : \quad SIG_A\{\!|n, m, B|\!\}$$

Figure 5.1: Challenge-response protocol as arrows-and-messages

5.2.1 MODELLING PROTOCOLS

In order to formally state and prove properties of security protocols, we first need to represent protocol parts as mathematical objects and define how they execute. The common informal arrows-and-messages notation (used, for example, in [50; 224]) is generally insufficient since it only presents the executions of the protocol that occur when there is no attack. One important part of security analysis involves understanding the way honest principals running a protocol will respond to messages from a malicious attacker. In addition, PCL requires more information about a protocol than the set of protocol executions obtained from honest and malicious parties; we need a high-level description of the program executed by each principal performing each protocol role so that we know not only which actions occur in a run and which do not, but why they occur.

Figure 5.1 shows a three-way signature based challenge-response protocol (CR) in the informal arrows-and-messages notation. The goal of the protocol – mutual authentication of two parties, is achieved by exchanging two fresh nonces m and n, and signatures over both nonces and the identity of the other party.

The roles of the same protocol are written as programs in Figure 5.2, writing $\hat{X}$ and $\hat{Y}$ for the principals executing roles **Init**$_{CR}$ and **Resp**$_{CR}$, respectively. We differentiate between *principals* (denoted by $\hat{X}$, $\hat{Y}$, ...) which correspond to protocol participants and may be involved in more than one execution of the protocol at any point and *threads* (denoted by X, Y, ...) which refer to a principal executing one particular session of the protocol. In this example, the protocol consists of two roles, the initiator role and the responder role. The sequence of actions in the initiator role is given by the program **Init**$_{CR}$ in Figure 5.2. In words, the actions of a principa executing the role **Init**$_{CR}$ are as follow as followss: generate a fresh random number; send a message with the random number to the peer $\hat{Y}$; receive a message with source address $\hat{Y}$; verify that the message contains $\hat{Y}$'s signature over the data in the expected format; and finally, send another message to $\hat{Y}$ with the initiator's signature over the nonce sent in the first message, the nonce received from $\hat{Y}$ and $\hat{Y}$'s identity. Formally, a *protocol* is a finite set of roles, one for each role of the protocol. In addition to the sequence of actions, the program for a role has static input and output parameters used when sequentially composing roles.

5.2.1.1 Protocol Programming Language

The protocol programming language is a conventional process calculus in the same vein as CCS, CSP, and their variants and descendants [133; 186]. However, since the protocols we focus on are a

$$\text{Init}_{CR} \equiv (\hat{Y})[\qquad\qquad\qquad \text{Resp}_{CR} \equiv ()[$$

$\text{Init}_{CR} \equiv (\hat{Y})[$	$\text{Resp}_{CR} \equiv ()[$
new m;	receive $\hat{X}, \hat{Y}, x$;
send $\hat{X}, \hat{Y}, m$;	new n;
receive $\hat{Y}, \hat{X}, y, s$;	$r := \text{sign}\,(n, x, \hat{X}), \hat{Y}$;
verify $s, (y, m, \hat{X}), \hat{Y}$;	send $\hat{Y}, \hat{X}, n, r$;
$r := \text{sign}\,(y, m, \hat{Y}), \hat{X}$;	receive $\hat{X}, \hat{Y}, t$;
send $\hat{X}, \hat{Y}, r$;	verify $t, (n, x, \hat{Y}), \hat{X}$;
$]_X ()$	$]_Y ()$

Figure 5.2: Roles of the Challenge-response protocol

concurrent composition of sequential roles, the process calculus is tailored to this form. The formalism was originally described in [100; 101] as *cord calculus*, a reference to the strand space formalism [108], which conveniently formalizes the practice of describing protocols by "arrows-and-messages", and displays the distributed traces of interacting processes. To represent the stores where the messages are to be received, we use process calculus variables and a substitution mechanism expressed by simple reaction rules, corresponding to the basic communication and computation operations. In comparison with conventional process calculus, we needed a mechanism for identifying the principal executing a sequence of actions, so that access to cryptographic keys could be identified and restricted. The resulting process calculus provides a protocol execution model, based on accepted concepts from process calculi, strand spaces and the chemical abstract machine. Its formal components are as follows.

Terms   A basic algebra of *terms t* is assumed to be given. As usual, they are built from constants c and variables x, by a given set of constructors p, which in this case includes tupling, public key encryption, and signature. We assume enough typing to distinguish the keys K from the principals $\hat{A}$, the nonces n and so on. Each type is given with enough variables.

As usual, computation is modelled as term evaluation. The closed terms, that can be completely evaluated, are the contents of the messages exchanged in protocols. The terms containing free variables (i.e. pointers and references) cannot be sent. An example term is $\hat{X}, \hat{Y}, m$ sent in the first message of the CR protocol (see Figure 5.1). It is important to note that $\hat{X}, \hat{Y}$ are parts of the message specifying intended sender and the recipient rather than parameters to the send action.

For technical purposes, we make a distinction between *basic terms u* which do not contain cryptographic operations explicitly (although, they may contain variables whose value is, for example, an encryption) and *terms t* which may contain cryptographic primitives.

Names, keys, sessions and threads   We use $\hat{A}, \hat{B}, \ldots$ as *names* for protocol participants. We will overload the notation and also use $\hat{A}, \hat{B}, \ldots$ as designation for public-private key pairs of the corresponding agents. A particular participant might be involved in more than one session at a time. For example, agent $\hat{A}$ involved in the CR protocol might be acting as an initiator in two sessions

(keys)	K	$::=$	k	basic key
			N	name
			$\overline{K}$	inverse key
(basic terms)	u	$::=$	x	basic term variable
			n	nonce
			N	name
			P	thread
			K	key
			u, u	tuple of basic terms
(terms)	t	$::=$	y	term variable
			u	basic term
			t, t	tuple of terms
			$ENC_K\{\!\mid\! t \!\mid\!\}$	term encrypted with key K
			$SIG_K\{\!\mid\! t \!\mid\!\}$	term signed with key $\overline{K}$

(actions)	a	$::=$	ϵ	the null action
			send u	send a term u
			receive x	receive term into variable x
			new x	generate new term x
			match u/u	match a term to a pattern
			$x :=$ sign u, K	sign the term u
			verify u, u, K	verify the signature
			$x :=$ enc u, K	encrypt the term u
			$x :=$ dec u, K	decrypt the term u
(strands)	S	$::=$	$[a; \ldots ; a]_P$	
(roles)	R	$::=$	$(\vec{x})S(\vec{t})$	

with agents $\hat{B}$ and $\hat{C}$ and as a responder in another parallel session with $\hat{D}$. For this reason, we will give names to sessions and use A to designate a particular *thread* being executed by $\hat{A}$.

Actions, strands and roles The set of actions contains nonce generation, encryption, decryption, signature generation and verification, pattern matching, testing and communication steps (sending and receiving). Pattern matching operator is used to construct and break tuples and perform equality checks. We will often omit the pattern matching operator and perform matching implicitly. For example, in the description of the CR protocol given in Figure 5.1 matching is implicitly done in the receive actions, if we were to completely write out actions there would be a `receive` x action followed by a `match` action analyzing the tuple, and performing the equality checks.

The list of actions will only contain basic terms which means that encryption cannot be performed implicitly; explicit `enc` action has to be used instead. For convenience, we assume that any variable will be assigned at most once, and the first occurrence of a particular variable has to be the assignment. Operational semantics of such single-assignment language will be significantly simpler as we can model the assignment with term substitution.

A *strand* is just a sequence of actions together with the designation of a thread performing the actions. A *role* is a strand with input and output interfaces used when performing sequential composition. All variables inside a role must be bound either by the input interface or by other actions. A *cord* is just a strand with no free variables, i.e. all ground terms are either constants or bound by a particular action.

5.2.1.2 Execution Model

A *protocol* Q is a set of roles $\{\rho_1, \rho_2, \ldots, \rho_k\}$, each executed by zero or more honest principals in any run of Q. Intuitively, these roles may correspond to the initiator, responder and the server, each specified by a sequence of actions to be executed in a single instance of a role. A protocol participant is called a *principal* and denoted by $\hat{A}, \hat{B}, \cdots$ etc. A single instance of a particular role executed by a principal will be called a *thread*. All threads of a single principal share static data such as long-term keys. As a notational convenience, we will use X to denote a thread of a principal $\hat{X}$. Protocols execute starting from an initial configuration containing threads of honest parties composed in parallel with an intruder process. The precise definition of protocol execution is based on process calculus operational semantics. We omit details here. The interested reader is referred to [89].

An *attack* is usually a process obtained by composing a protocol with another process, in such a way that the resulting runs, projected to the protocol roles, do not satisfy the protocol requirements. An *attacker*, or *intruder*, is a set of threads sharing all data in an attack, and playing roles in one or more protocol sessions. The actions available for building the intruder roles usually include receiving and sending messages, decomposing them into parts, decrypting them by known keys, storing data, and even generating new data. This is the standard "Dolev-Yao model", which appears to have developed from positions taken by Needham and Schroeder [201] and a model presented by Dolev and Yao [97].

5.2.2 PROTOCOL LOGIC

5.2.2.1 Syntax

The formulas of PCL are given by the grammar in Table 5.3, where S may be any strand. Here, t and P denote a term and a *thread*, respectively. We use ϕ and ψ to indicate predicate formulas, and m to indicate a generic term we call a "message". A message has the form (source, destination,

Action formulas

$$a \quad ::= \quad \mathsf{Send}(P, t) \mid \mathsf{Receive}(P, t) \mid \mathsf{New}(P, t) \mid \mathsf{Encrypt}(P, t) \mid$$
$$\mathsf{Decrypt}(P, t) \mid \mathsf{Sign}(P, t) \mid \mathsf{Verify}(P, t)$$

Formulas

$$\phi \quad ::= \quad a \mid a < a \mid \mathsf{Has}(P, t) \mid \mathsf{Fresh}(P, t) \mid \mathsf{Gen}(P, t) \mid \mathsf{FirstSend}(P, t, t) \mid$$
$$\mathsf{Honest}(N) \mid t = t \mid \mathsf{Contains}(t, t) \mid \phi \wedge \phi \mid \neg \phi \mid \exists x. \phi \mid \mathsf{Start}(P)$$

Modal formulas

$$\Psi \quad ::= \quad \phi \, S \, \phi$$

protocol-identifier, content), giving each message source and destination fields and a unique protocol identifier in addition to the message contents. The source field of a message may not identify the actual sender of the message since the intruder can spoof the source address. Similarly, the principal identified by the destination field may not receive the message since the intruder can intercept messages. Nonetheless, the source and destination fields in the message may be useful for stating and proving authentication properties while the protocol-identifier is useful for proving properties of protocols.

Most protocol proofs use formulas of the form $\theta[P]_X \phi$, which means that after actions P are executed in thread X, starting from a state where formula θ is true, formula ϕ is true about the resulting state of X. Here are the informal interpretations of the predicates, with the precise semantics discussed elsewhere [88].

Action formulas Action formulas are used to state that particular actions have been performed by various threads. The formula $\mathsf{Send}(X, m)$ means that principal $\hat{X}$ sent a message m in the thread X. Predicates Receive, Encrypt, Sign, $\cdots$, etc., are similarly used to state that the corresponding actions have been performed. Action predicates are crucial in modelling authentication properties of the protocol. In PCL, the fact that $\hat{A}$ has authenticated $\hat{B}$ will be described by saying that $\hat{B}$ must have performed certain actions prescribed by the protocol.

Knowledge The formula $\mathsf{Has}(X, x)$ means that principal $\hat{X}$ possesses information x in the thread X. This is "possess" in the limited sense of having either generated the data or received it in the clear or received it under encryption where the decryption key is known. The formula $\mathsf{Fresh}(X, t)$ means that the term t generated in X is "fresh" in the sense that no one else has seen any term containing t as a subterm. Typically, a fresh term will be a nonce and freshness will be used to reason about the temporal ordering of actions in runs of a protocol. The formula $\mathsf{Gen}(X, t)$ means that the term t originated in the thread X in the sense that it was "fresh" in X at some point. The formula $\mathsf{Contains}(t_1, t_2)$ means that the term t_1 contains term t_2 as a subterm. Predicate Has can be used to model secrecy properties; for example, a fact that a term t is a shared secret between threads X and Y is captured by the logical formula $\forall Z. \mathsf{Has}(Z, t) \supset (Z = X \vee Z = Y)$.

Temporal ordering   The formula $\mathsf{Start}(X)$ means that the thread X did not execute any actions in the past. The formula, $a_1 < a_2$ means that both actions a_1 and a_2 happened in the run and, moreover, that the action a_2 happened after the action a_1. Note that actions may not be unique. For example, a thread X might have received the same term multiple times, temporal ordering operator only states that *some* two actions a_1 and a_2 have happened in that order. The formula $\mathsf{FirstSend}(P, t, t')$ means that the thread P has send a term t (possibly as a part of some bigger message) and that the first such occurrence was an action when P send the message t'. Temporal ordering relation can be used to strengthen the authentication properties by imposing ordering between actions of different participants.

Honesty   The formula $\mathsf{Honest}(\hat{X})$ means the actions of principal $\hat{X}$ in the current run are precisely an interleaving of initial segments of traces of a set of roles of the protocol. In other words, each thread X of principal $\hat{X}$ assumes a particular role of the protocol and does exactly the actions prescribed by that role.

Modal Formulas   Modal formulas attach assertions – *preconditions* and *postconditions* – to programs. Informally, formula of the form $\theta[P]_X\phi$ means that after actions P are executed in thread X, starting from a state where formula θ is true, formula ϕ is true about the resulting state of X.

5.2.2.2 Semantics
A formula may be true or false at a run of a protocol. More precisely, the main semantic relation, $\mathcal{Q}, R \models \phi$, may be read, "formula ϕ holds for run R of protocol $\mathcal{Q}$." In this relation, R may be a complete run, with all sessions that are started in the run completed, or an incomplete run with some principals waiting for additional messages to complete one or more sessions. If $\mathcal{Q}$ is a protocol, then let $\bar{\mathcal{Q}}$ be the set of all initial configurations of protocol $\mathcal{Q}$, each including a possible intruder cord. Let $\mathsf{Runs}(\mathcal{Q})$ be the set of all runs of protocol $\mathcal{Q}$ with intruder, each beginning from an initial configuration in $\bar{\mathcal{Q}}$ sequence of reaction steps within a cord space. If ϕ has free variables, then $\mathcal{Q}, R \models \phi$ if we have $\mathcal{Q}, R \models \sigma\phi$ for all substitutions σ that eliminate all the free variables in ϕ. We write $\mathcal{Q} \models \phi$ if $\mathcal{Q}, R \models \phi$ for all $R \in \mathsf{Runs}(\mathcal{Q})$. The full definition of the semantics is given in [88].

5.2.3   PROOF SYSTEM
The proof system combines a complete axiom system for first-order logic (not listed since any axiomatization will do), together with axioms and proof rules for protocol actions, temporal reasoning, and a specialized form of invariance rule.

5.2.3.1 Axioms for Protocol Actions
Axioms for protocol actions state properties that hold in a state as a result of executing certain actions (or not executing certain actions). We use a in the axioms to denote any one of the actions and a to denote the corresponding predicate in the logic. $\top$ denotes the boolean value *true*. Axiom **AA1** states that if a principal has executed an action in some role, then the corresponding predicate asserting

that the action had occurred in the past is true while **AA2** states that at the start of a thread any action predicate applied to the thread is false. Axiom **AA3** states that the predicate asserting thread X has not sent the term t remains false after any action that does not send a term that unifies with t, if it is false before the action. **AA4** states that after thread X does actions $a, \cdots, b$ in sequence, the action predicates, a and b, corresponding to the actions, are temporally ordered in the same sequence.

AA1	$\top[a]_X$ a
AA2	$\text{Start}(X)[\ ]_X \ \neg a(X)$
AA3	$\neg\text{Send}(X, t)[b]_X \neg\text{Send}(X, t)$
	if $\sigma\,\text{Send}(X, t) \neq \sigma\,b$ for all substitutions σ
AA4	$\top[a; \cdots ; b]_X\, a < b$

The following axioms deal with properties of freshly generated nonces. Axiom **AN1** states that a particular nonce is generated by a unique thread. If thread X generates a new value n and does no further actions, then axiom **AN2** says that no one else knows n, and axiom **AN3** says that n is fresh, and axiom **AN4** says that X is the originating thread of nonce n.

AN1	$\text{New}(X, x) \wedge \text{New}(Y, x) \supset X = Y$
AN2	$\top[\texttt{new } x]_X\, \text{Has}(Y, x) \supset (Y = X)$
AN3	$\top[\texttt{new } x]_X\, \text{Fresh}(X, x)$
AN4	$\text{Fresh}(X, x) \supset \text{Gen}(X, x)$

5.2.3.2 Possession Axioms

The possession axioms characterize the terms that a principal can derive if it possesses certain other terms. **ORIG** and **REC** state, respectively, that a principal possesses a term if she freshly generated it (a nonce) or if she received it in some message. **TUP** and **ENC** enable construction of tuples and encrypted terms if the parts are known. **PROJ** and **DEC** allow decomposition of a tuple into its components and decryption of an encrypted term if the key is known.

ORIG	$\text{New}(X, x) \supset \text{Has}(X, x)$
REC	$\text{Receive}(X, x) \supset \text{Has}(X, x)$
TUP	$\text{Has}(X, x) \wedge \text{Has}(X, y) \supset \text{Has}(X, (x, y))$
ENC	$\text{Has}(X, x) \wedge \text{Has}(X, K) \supset \text{Has}(X, ENC_K \{\!|x|\!\})$
PROJ	$\text{Has}(X, (x, y)) \supset \text{Has}(X, x) \wedge \text{Has}(X, y)$
DEC	$\text{Has}(X, ENC_K \{\!|x|\!\}) \wedge \text{Has}(X, K) \supset \text{Has}(X, x)$

Axioms **AR1, AR2** and **AR3** are used to model obtaining information about structure of terms as they are being parsed. They allow us to plug in appropriate substitutions obtained by matching, signature verification and decryption actions to terms inside the action predicate a.

$$\textbf{AR1} \qquad \mathsf{a}(x)[\mathtt{match}\ q(x)/q(t)]_X\ \mathsf{a}(t)$$
$$\textbf{AR2} \qquad \mathsf{a}(x)[\mathtt{verify}\ x, t, K]_X\ \mathsf{a}(SIG_K\{\!|t|\!\})$$
$$\textbf{AR3} \qquad \mathsf{a}(x)[y := \mathtt{dec}\ x, K]_X\ \mathsf{a}(ENC_K\{\!|y|\!\})$$

5.2.3.3 Encryption and Signature

The next two axioms are aimed at capturing the black-box model of encryption and signature. Axiom **VER** refers to the unforgeability of signatures while axiom **SEC** stipulates the need to possess the private key in order to decrypt a message encrypted with the corresponding public key.

$$\textbf{SEC} \qquad \mathsf{Honest}(\hat{X}) \wedge \mathsf{Decrypt}(Y, ENC_{\hat{X}}\{\!|x|\!\}) \supset (\hat{Y} = \hat{X})$$
$$\textbf{VER} \qquad \mathsf{Honest}(\hat{X}) \wedge \mathsf{Verify}(Y, SIG_{\hat{X}}\{\!|x|\!\}) \wedge \hat{X} \neq \hat{Y} \supset$$
$$\exists X.\mathsf{Send}(X, m) \wedge \mathsf{Contains}(m, SIG_{\hat{X}}\{\!|x|\!\})$$

5.2.3.4 Generic Rules

These are generic Floyd-Hoare style rules for reasoning about program pre-conditions and post-conditions. For example, the generalization rule **G4** says that if ϕ is a valid formula (it holds in all runs of all protocols) then it can be used in a postcondition of any modal form.

$$\frac{\theta[P]_X\phi \qquad \theta[P]_X\psi}{\theta[P]_X\phi \wedge \psi}\ \textbf{G1} \qquad\qquad \frac{\theta[P]_X\psi \qquad \phi[P]_X\psi}{\theta \vee \phi[P]_X\psi}\ \textbf{G2}$$

$$\frac{\theta' \supset \theta \qquad \theta[P]_X\phi \qquad \phi \supset \phi'}{\theta'[P]_X\phi'}\ \textbf{G3} \qquad\qquad \frac{\phi}{\theta[P]_X\phi}\ \textbf{G4}$$

5.2.3.5 Sequencing Rule

Sequencing rule gives us a way of sequentially composing two cords P and P', when post-condition of P, matches the pre-condition of P'.

$$\frac{\phi_1[P]_X\phi_2 \qquad \phi_2[P']_X\phi_3}{\phi_1[PP']_X\phi_3}\ \textbf{S1}$$

5.2.3.6 Preservation Axioms

The following axioms state that the truth of certain predicates continue to hold after further actions. **P1** states this for the predicates {Has, FirstSend, a} whereas **P2** states that freshness of a term holds across actions that do not send out some term containing it.

$$\textbf{P1} \qquad \mathsf{Persist}(X, t)[a]_X\mathsf{Persist}(X, t)$$
$$\text{for } \mathsf{Persist} \in \{\mathsf{Has}, \mathsf{FirstSend}, \mathsf{a}, \mathsf{Gen}\}.$$
$$\textbf{P2} \qquad \mathsf{Fresh}(X, t)[a]_X\mathsf{Fresh}(X, t)$$
$$\text{where } t \not\subseteq a.$$

5.2.3.7 Axioms and Rules for Temporal Ordering

The next two axioms give us a way of deducing temporal ordering between actions of different threads. Informally, $\text{FirstSend}(X, t, t')$ says that a thread X generated a fresh term t and sent it out first in message t'. This refers to the first such send event and is formally captured by axiom **FS1**. Axiom **FS2** lets us reason that if a thread Y does some action with a term t'', which contains a term t, first sent inside a term t' by a thread X as a subterm, then that send must have occurred before Y's action.

FS1 $\quad\quad$ $\text{Fresh}(X, t)[\text{send } t']_X \text{FirstSend}(X, t, t')$
$\quad\quad\quad\quad\quad$ where $t \subseteq t'$.

FS2 $\quad\quad$ $\text{FirstSend}(X, t, t') \wedge \text{a}(Y, t'') \supset \text{Send}(X, t') < \text{a}(Y, t'')$
$\quad\quad\quad\quad\quad$ where $X \neq Y$ and $t \subseteq t''$.

5.2.3.8 The Honesty Rule

The honesty rule is an invariance rule for proving properties about the actions of principals that execute roles of a protocol, similar in spirit to the basic invariance rule of LTL [171] and invariance rules in other logics of programs. The honesty rule is often used to combine facts about one role with inferred actions of other roles. For example, suppose Alice receives a signed response from a message sent to Bob. Alice may use facts about Bob's role to infer that Bob must have performed certain actions before sending his reply. This form of reasoning may be sound if Bob is honest, since honest, by definition in this framework, means "follows one or more roles of the protocol." The assumption that Bob is honest is essential because the intruder may perform arbitrary actions with any key that has been compromised. To a first approximation, the honesty rule says that if a property holds before each role starts, and the property is preserved by any sequence of actions that an honest principal may perform, then the property holds for every honest principal.

Recall that a protocol $\mathcal{Q}$ is a set of roles $\{\rho_1, \rho_2, \ldots, \rho_k\}$, each executed by zero or more honest principals in any run of $\mathcal{Q}$. A sequence P of actions is a *basic sequence* of role ρ, written $P \in BS(\rho)$, if P is a contiguous subsequence of ρ such that either (i) P starts at the beginning of ρ and ends with the last action before the first receive, or (ii) P starts with a receive action and continues up to the last action before the next receive, or (iii) P starts with the last receive action of the role and continues through the end of the role. In the syntactic presentation below, we use the notation $\forall \rho \in \mathcal{Q}. \forall P \in BS(\rho). \phi[P]_X \phi$ to denote a finite set of formulas of the form $\phi[P]_X \phi$ - one for each basic sequence P in the protocol. The quantifiers $\forall \rho \in \mathcal{Q}$ and $\forall P \in BS(\rho)$ are not part of the syntax of PCL, but are meta-notation used to state this rule schema.

$$\frac{\text{Start}(X)[\ ]_X \phi \quad\quad \forall \rho \in \mathcal{Q}. \forall P \in BS(\rho). \phi [P]_X \phi}{\text{Honest}(\hat{X}) \supset \phi} \textbf{HON}_\mathbf{Q} \quad \begin{array}{l} \text{no free variable in } \phi \text{ ex-} \\ \text{cept } X \text{ bound in } [P]_X \end{array}$$

5.2.3.9   Soundness

The soundness theorem for this proof system is proved, by induction on the length of proofs. We write $\Gamma \vdash \gamma$ if γ is provable from the formulas in Γ and any axiom or inference rule of the proof system except the honesty rule ($\mathbf{HON_Q}$ for any protocol $\mathcal{Q}$). We write $\Gamma \vdash_Q \gamma$ if γ is provable from the formulas in Γ, the basic axioms and inference rules of the proof system and the honesty rule for protocol $\mathcal{Q}$ (i.e., $\mathbf{HON_Q}$ but not $\mathbf{HON_{Q'}}$ for any $\mathcal{Q'} \neq \mathcal{Q}$). Here γ is either a modal formula or a basic formula (i.e., of the syntactic form Ψ or ϕ in Table 5.3).

Theorem 5.1   *If* $\Gamma \vdash_Q \gamma$, *then* $\Gamma \models_Q \gamma$. *Furthermore, if* $\Gamma \vdash \gamma$, *then* $\Gamma \models \gamma$.

5.2.4   EXAMPLE

In this section, we use the protocol logic to formally prove the authentication property of the three-way signature based challenge-response protocol (CR) described in Section 5.2.1. Our formulation of authentication is based on the concept of *matching conversations* [38] and is similar to the idea of proving authentication using *correspondence assertions* [255]. The same basic idea is also presented in [95] where it is referred to as *matching records of runs*. Simply put, it requires that whenever Alice and Bob accept each other's identities at the end of a run, their records of the run *match*, i.e., each message that Alice sent was received by Bob and vice versa, each send event happened before the corresponding receive event, and, moreover, the messages sent by each principal appear in the same order in both the records. Here we prove the authentication property only for the initiator in the protocol; the security proof for the responder is similar.

Weak authentication   First, we show a weaker authentication property. If Alice has completed the initiator role of the protocol, apparently with Bob then Bob was involved in the protocol – he received the first message and sent out the corresponding second message. The formal property proved about the initiator role is

$$\vdash_{Q_{CR}} \top [\mathbf{Init_{CR}}]_X \text{Honest}(\hat{Y}) \wedge \hat{Y} \neq \hat{X} \supset \phi_{weak-auth}.$$

The actions in the modal formula are the actions of the initiator role of CR, given in Section 5.2.1. The precondition imposes constraints on the free variables. In this example, the precondition is simply "true". The postcondition captures the security property that is guaranteed by executing the actions starting from a state where the precondition holds. In this specific example, the postcondition is a formula capturing the notion of weak authentication. Intuitively, this formula means that after executing the actions in the initiator role purportedly with $\hat{Y}$, $\hat{X}$ is guaranteed that $\hat{Y}$ was involved in the protocol at some point (purportedly with $\hat{X}$), provided that $\hat{Y}$ is honest (meaning that she always faithfully executes some role of the CR protocol and does not, for example, send out her private keys).

$$\phi_{weak-auth} \equiv \exists Y. \, (\text{Receive}(Y, (\hat{X}, \hat{Y}, m)) < \text{Send}(Y, (\hat{Y}, \hat{X}, y, SIG_{\hat{Y}}\{|y, m, \hat{X}|\})))$$

$$\mathbf{AA1} \qquad \top[\text{verify } s, (y, m, \hat{X}), \hat{Y}]_X \text{Verify}(X, SIG_{\hat{Y}}\{\!|y, m, \hat{X}|\!\}) \tag{5.1}$$

$$\mathbf{5.1, P1, SEQ} \qquad \top[\mathbf{Init}_{CR}]_X \text{Verify}(X, SIG_{\hat{Y}}\{\!|y, m, \hat{X}|\!\}) \tag{5.2}$$

$$\mathbf{5.2, VER} \qquad \top[\mathbf{Init}_{CR}]_X \exists Y, t.\ \text{Send}(Y, t) \wedge \text{Contains}(t, SIG_{\hat{Y}}\{\!|y, m, \hat{X}|\!\}) \tag{5.3}$$

$$\mathbf{HON}_{Q_{CR}} \qquad (\text{Honest}(\hat{Y}) \wedge \text{Send}(Y, t) \wedge \text{Contains}(t, SIG_{\hat{Y}}\{\!|y, m, \hat{X}|\!\})) \supset \tag{5.4}$$
$$(\text{New}(Y, m) \vee$$
$$(\text{Receive}(Y, (\hat{X}, \hat{Y}, m)) < \text{Send}(Y, (\hat{Y}, \hat{X}, y, SIG_{\hat{Y}}\{\!|y, m, \hat{X}|\!\}))))$$

$$\mathbf{5.3, 5.4} \qquad \top[\mathbf{Init}_{CR}]_X \text{Honest}(\hat{Y}) \supset (\exists Y.\ \text{New}(Y, m) \vee \tag{5.5}$$
$$(\text{Receive}(Y, (\hat{X}, \hat{Y}, m)) < \text{Send}(Y, (\hat{Y}, \hat{X}, y, SIG_{\hat{Y}}\{\!|y, m, \hat{X}|\!\}))))$$

$$\mathbf{AA1} \qquad \top[\text{new } m]_X \text{New}(X, y) \tag{5.6}$$

$$\mathbf{5.6, P1, SEQ} \qquad \top[\mathbf{Init}_{CR}]_X \text{New}(X, y) \tag{5.7}$$

$$\mathbf{5.5, 5.7, AN1} \qquad \top[\mathbf{Init}_{CR}]_X \text{Honest}(\hat{Y}) \wedge \hat{Y} \neq \hat{X} \supset (\exists Y. \tag{5.8}$$
$$\text{Receive}(Y, (\hat{X}, \hat{Y}, m)) < \text{Send}(Y, (\hat{Y}, \hat{X}, y, SIG_{\hat{Y}}\{\!|y, m, \hat{X}|\!\})))$$

A formal proof of the weak authentication property for the initiator guaranteed by executing the CR protocol is presented in Table 5.4. First order logic reasoning steps as well as applications of the generic rules are omitted for clarity. Details for the application of the honesty rule are postponed until later in this section. The formal proof naturally breaks down into three parts:

- Lines (5.1)–(5.3) assert what actions were executed by Alice in the initiator role. Specifically, in this part of the proof, it is proved that Alice has received and verified Bob's signature $SIG_{\hat{Y}}\{\!|y, m, \hat{X}|\!\}$. We then use the fact that the signatures of honest parties are unforgeable (axiom **VER**), to conclude that Bob must have sent out some message containing his signature.

- In lines (5.4)–(5.5), the honesty rule is used to infer that whenever Bob generates a signature of this form, he has either generated the nonce m (acting as an initiator), or he sent it to Alice as part of the second message of the protocol and must have previously received the first message from Alice (acting as a responder).

- Finally, in lines (5.6)–(5.8), we reason again about actions executed by Alice in order to deduce that the nonce m could not have been created by Bob. Therefore, combining the assertions, we show that the weak authentication property holds: If Alice has completed the protocol as an initiator, apparently with Bob, then Bob must have received the first message (apparently from Alice) and sent the second message to Alice.

Strong authentication To obtain the stronger authentication property, we need to assert temporal ordering between actions of Alice and Bob. As mentioned before, the final authentication property should state that: each message $\hat{X}$ sent was received by $\hat{Y}$ and vice versa, each send event happened before the corresponding receive event, and moreover the messages sent by each principal ($\hat{X}$ or

$$\textbf{AN3} \qquad \top[\text{new } m]_X \text{Fresh}(X, m) \tag{5.9}$$

$$\textbf{FS1} \qquad \text{Fresh}(X, m)[\text{send } \hat{X}, \hat{Y}, m]_X \text{FirstSend}(X, m, (\hat{X}, \hat{Y}, m)) \tag{5.10}$$

$$\text{5.9, 5.10, } \textbf{SEQ, P1} \qquad \top[\textbf{Init}_{CR}]_X \text{FirstSend}(X, m, (\hat{X}, \hat{Y}, m)) \tag{5.11}$$

$$\text{5.11, } \textbf{FS2} \qquad \top[\textbf{Init}_{CR}]_X \text{Receive}(Y, (\hat{X}, \hat{Y}, m)) \wedge \hat{Y} \neq \hat{X} \supset \tag{5.12}$$
$$\text{Send}(X, (\hat{X}, \hat{Y}, m)) < \text{Receive}(Y, (\hat{X}, \hat{Y}, m))$$

$$\textbf{HON}_{\textbf{Q}_{CR}} \qquad (\text{Honest}(\hat{Y}) \wedge \text{Receive}(Y, (\hat{X}, \hat{Y}, m)) \wedge \tag{5.13}$$
$$\text{Send}(Y, (\hat{Y}, \hat{X}, y, SIG_{\hat{Y}}\{\!|y, m, \hat{X}|\!\}))) \supset$$
$$\text{FirstSend}(Y, y, (\hat{Y}, \hat{X}, y, SIG_{\hat{Y}}\{\!|y, m, \hat{X}|\!\}))$$

$$\textbf{AA1, AR2, SEQ} \qquad \top[\textbf{Init}_{CR}]_X \text{Receive}(X, (\hat{Y}, \hat{X}, y, SIG_{\hat{Y}}\{\!|y, m, \hat{X}|\!\})) \tag{5.14}$$

$$\text{5.13, 5.14, } \textbf{FS2} \qquad \top[\textbf{Init}_{CR}]_X \text{Honest}(\hat{Y}) \wedge \hat{Y} \neq \hat{X} \wedge \tag{5.15}$$
$$\text{Receive}(Y, (\hat{X}, \hat{Y}, m)) \wedge$$
$$\text{Send}(Y, (\hat{Y}, \hat{X}, y, SIG_{\hat{Y}}\{\!|y, m, \hat{X}|\!\})) \supset$$
$$\text{Send}(Y, (\hat{Y}, \hat{X}, y, SIG_{\hat{Y}}\{\!|y, m, \hat{X}|\!\})) <$$
$$\text{Receive}(X, (\hat{Y}, \hat{X}, y, SIG_{\hat{Y}}\{\!|y, m, \hat{X}|\!\}))$$

$$\textbf{AA4, P1} \qquad \top[\textbf{Init}_{CR}]_X \text{Receive}(X, (\hat{Y}, \hat{X}, y, SIG_{\hat{Y}}\{\!|y, m, \hat{X}|\!\})) < \tag{5.16}$$
$$\text{Send}(X, (\hat{X}, \hat{Y}, SIG_{\hat{X}}\{\!|y, m, \hat{Y}|\!\}))$$

$$\text{5.8, 5.12, 5.15, 5.16} \qquad \top[\textbf{Init}_{CR}]_X \text{Honest}(\hat{Y}) \wedge \hat{Y} \neq \hat{X} \supset \phi_{auth} \tag{5.17}$$

$\hat{Y}$) appear in the same order in both the records. Similarly as before, the formal property proved about the initiator role is $\vdash_{Q_{CR}} \top[\textbf{Init}_{CR}]_X \text{Honest}(\hat{Y}) \wedge \hat{Y} \neq \hat{X} \supset \phi_{auth}$, but ϕ_{auth} now models the stronger property:

$$\phi_{auth} \equiv \exists Y. ((\text{Send}(X, msg_1) < \text{Receive}(Y, msg_1)) \wedge$$
$$(\text{Receive}(Y, msg_1) < \text{Send}(Y, msg_2)) \wedge$$
$$(\text{Send}(Y, msg_2) < \text{Receive}(X, msg_2)) \wedge$$
$$(\text{Receive}(X, msg_2) < \text{Send}(X, msg_3)))$$

Here, we are using msg_1, msg_2 and msg_3 as shortcuts for the corresponding messages in the protocol: $msg_1 \equiv (\hat{X}, \hat{Y}, m)$, $msg_2 \equiv (\hat{Y}, \hat{X}, y, SIG_{\hat{Y}}\{\!|y, m, \hat{X}|\!\})$, $msg_3 \equiv (\hat{X}, \hat{Y}, SIG_{\hat{X}}\{\!|y, m, \hat{Y}|\!\})$. Note that we cannot deduce that the responder Y has received the third message as that property does not necessarily hold from the point of view of the initiator.

A formal proof of the strong authentication property for the initiator guaranteed by executing the CR protocol is presented in Table 5.5. Again, the formal proof naturally breaks down into three parts:

- Lines (5.9)–(5.12) reason about actions executed by Alice in the initiator role. Specifically, it is proved that the first occurrence of the nonce m on the network is in the first message sent by Alice. Hence, all actions involving that nonce must happen after that send action.

- In lines (5.13)–(5.15), the honesty rule is used to infer the symmetrical property about Bob's nonce y. Hence, all actions involving that nonce must happen after the send action by Bob in the second step of the protocol.

- In line (5.16), we reason from Alice's actions that she sent out the third message after receiving the second message.

- Finally, in line (5.17), the weak authentication property already proved is combined with the newly established temporal assertions to infer the final strong authentication property.

The proofs together are an instance of a general method for proving authentication results in PCL. In proving that Alice, after executing the initiator role of a protocol purportedly with Bob, is indeed assured that she communicated with Bob, we usually follow these 3 steps:

1. Prove the order in which Alice executed her send-receive actions. This is done by examining the actions in Alice's role.

2. Assuming Bob is honest, infer the order in which Bob carried out his send-receive actions. This is done in two steps. First, use properties of cryptographic primitives (like signing and encryption) to conclude that only Bob could have executed a certain action (e.g., generate his signature). Then use the honesty rule to establish a causal relationship between that identifying action and other actions that Bob always does whenever he executes that action (e.g, send msg_2 to Alice after having received msg_1 from her).

3. Finally, use the temporal ordering rules to establish an ordering between the send-receive actions of Alice and Bob. The causal ordering between messages sent by the peers is typically established by exploiting the fact that messages contain fresh data.

Proofs in the logic are therefore quite insightful. The proof structure often follows a natural language argument, similar to one that a protocol designer might use to convince herself of the correctness of a protocol.

Invariants   In both proofs, the honesty rule is used to deduce that the other party in the protocol has performed certain actions or not. Formulas proved by the application of the honesty rule are called *invariants*. This proof uses two invariants $\mathsf{Honest}(\hat{Y}) \supset \gamma_1$ and $\mathsf{Honest}(\hat{Y}) \supset \gamma_2$ where γ_1 and

γ_2 are given by the following:

$$
\begin{aligned}
\gamma_1 \;\equiv\; & \mathsf{Send}(Y, t) \wedge \mathsf{Contains}(t, SIG_{\hat{Y}}\{\!|\, y, m, \hat{X}\,|\!\}) \supset \\
& (\mathsf{Gen}(Y, m) \vee \\
& (\mathsf{Receive}(Y, (\hat{X}, \hat{Y}, m)) < \mathsf{Send}(Y, (\hat{Y}, \hat{X}, y, SIG_{\hat{Y}}\{\!|\, y, m, \hat{X}\,|\!\})))) \\
\gamma_2 \;\equiv\; & (\mathsf{Receive}(Y, (\hat{X}, \hat{Y}, m)) \wedge \mathsf{Send}(Y, (\hat{Y}, \hat{X}, y, SIG_{\hat{Y}}\{\!|\, y, m, \hat{X}\,|\!\}))) \supset \\
& \mathsf{FirstSend}(Y, y, (\hat{Y}, \hat{X}, y, SIG_{\hat{Y}}\{\!|\, y, m, \hat{X}\,|\!\}))
\end{aligned}
$$

As described in Section 5.2.3.8, the honesty rule depends on the protocol being analyzed. Recall that the protocol Q_{CR} is just a set of roles $Q_{CR} = \{\mathbf{Init_{CR}}, \mathbf{Resp_{CR}}\}$ each specifying a sequence of actions to be executed. The set of basic sequences of protocol Q_{CR} is given below.

$$
\begin{aligned}
\mathbf{BS_1} \;\equiv\;& [\texttt{new } m;\, \texttt{send } \hat{X}, \hat{Y}, m;\,]_X \\
\mathbf{BS_2} \;\equiv\;& [\texttt{receive } \hat{Y}, \hat{X}, y, s;\, \texttt{verify } s, (y, m, \hat{X}), \hat{Y};\, \\
& \quad r := \texttt{sign } (y, m, \hat{Y}), \hat{X};\, \texttt{send } \hat{X}, \hat{Y}, r;\,]_X \\
\mathbf{BS_3} \;\equiv\;& [\texttt{receive } \hat{X}, \hat{Y}, x;\, \texttt{new } n;\, r := \texttt{sign } (n, x, \hat{X}), \hat{Y};\, \texttt{send } \hat{Y}, \hat{X}, n, r;\,]_Y \\
\mathbf{BS_4} \;\equiv\;& [\texttt{receive } \hat{X}, \hat{Y}, t;\, \texttt{verify } t, (n, x, \hat{Y}), \hat{X};\,]_Y
\end{aligned}
$$

Therefore, to apply the honesty rule, we need to show that the invariants (γ_1, γ_2) are preserved by all the basic sequences $(\mathbf{BS}_1, \ldots, \mathbf{BS}_4)$. These proofs are straightforward, and we omit them here.

5.3 OTHER PROTOCOL ANALYSIS APPROACHES

In this section, we summarize and provide pointers to some of the other representative methods and tools for security protocol analysis based on model checking, theorem-proving, symbolic search, process calculi, and multiset rewriting.

Model-checking is a technique for automatically analyzing properties of systems modeled as finite state transition systems. A number of model-checking approaches have been developed for analyzing security protocols [18; 24; 46; 70; 168; 185; 191; 224; 236; 244]. The first work in this area was based on the CSP process calculus and the associated FDR model checker [224]. Lowe identified the well-known attack on the Needham-Scroeder public key protocol using this method [168]. Another successful model-checking method for security protocols [191] uses Murφ [96], a generic finite state model-checker. This method has been used to successfully analyze industrial security protocols including SSL [194] and a component of the IEEE 802.11i protocol suite [127]. It has also been used by students to carry out realistic protocol analysis case studies in graduate-level security courses. Murφ as well as the model for SSL are available online [1; 2] and could be used for course projects. The AVISPA project [24] has recently developed a suite of protocol analysis tools, including a state-of-the-art model checker, OFMC [32].

Paulson developed a semi-automated theorem-proving method for proving secrecy and authentication properties of network protocols [208] in which the operational semantics of all agents in the network including the attacker is modeled using a series of inductive definitions. This method was successfully used to analyze several industrial protocols including TLS, Kerberos, and parts of SET [34; 35; 209]. The case studies were carried out using the generic theorem-prover Isabelle. Other theorem-proving approaches to protocol verification include Cohen's work on first-order reasoning about protocols [71; 72].

Meadows developed the NRL Protocol Analyzer (NPA) [178] and, with co-authors, has successfully applied the tool to formally verify and/or identify security vulnerabilities in a number of industrial protocols. The case studies include the IETF standards IKE [179] and GDOI [182]. NPA uses a combination of symbolic reachability analysis using narrowing and techniques for reducing the size of the search space. Specifically, it uses inductively defined conditions to specify states that are unreachable by an attacker thereby succeeding (in many cases) to reduce an analysis of a protocol with an unbounded number of sessions to a finite state search problem. One powerful feature of NPA is that it can be used to reason about the algebraic properties of functions used, for example, for encryption and decryption [177].

A number of process calculi have also been developed for reasoning about security protocols. Abadi and Gordon developed the spi-calculus [11], an extension of the pi-calculus [187; 188] with cryptographic primitives. Protocols are represented as processes in the spi-calculus. The channel abstraction, inherited from the pi-calculus, is useful for modeling communication between parties and the scoping rules for channels provide a mechanism to express that the adversary cannot access information communicated on certain channels. Security conditions are stated using specialized forms of process equivalence. The idea here is that the protocol whose security we are trying to evaluate should be "equivalent" to an ideal protocol which is "obviously" secure (e.g. because it uses private channels to communicate secrets). A related approach is taken in the applied pi-calculus of Abadi and Fournet [10] which has been used to analyze the JFK protocol [9]; part of this analysis was carried out using ProVerif, Blanchet's logic programming-based automatic protocol analysis tool [39]. Another project in this space is CryptoSPA [110].

A specialized model for representing security protocol executions and associated proof techniques for authentication and secrecy properties are developed in the strand spaces project [108]. The model captures causal interaction among multiple protocol participants and an adversary. A class of proof techniques called "authentication tests" were developed for establishing authentication properties over the semantic model. An early implementation of this analysis method is Song's tool, Athena [244].

Another line of work examines the decidability and complexity of the protocol analysis problem. A multiset rewriting-based framework (MSR) [102] was developed by Durgin *et al* to formally represent protocols and their execution. Decidability and complexity of the protocol insecurity problem was studied in the MSR model for various execution scenarios, e.g. unbounded number of sessions, finite number of sessions, etc. (see [102] for details). Rusinowitch and Turuani improved one aspect of these results by showing that the protocol insecurity problem for a finite number of sessions remains NP-complete even when the intruder is allowed to construct terms of unbounded size [223]. These results have been extended to allow for equational theories arising from the algebraic properties of functions such as XOR and Diffie-Hellman primitives used in cryptographic protocols (see [75] for a recent survey of these results).

5.4   RECENT ADVANCES

In this section, we report on progress on two significant protocol analysis problems that have been addressed in recent work (see [180] for these and other major open problems in this area). Progress on methods for *secure protocol composition* is reported in Section 5.4.1 while *cryptographically sound protocol analysis methods* are described in Section 5.4.2.

5.4.1   SECURE COMPOSITION

Early work on the protocol composition problem concentrated on designing protocols that would be guaranteed to compose with any other protocol. This led to rather stringent constraints on protocols: in essence, they required the fail-stop property [117] or something very similar to it [130]. Since real-world protocols are not designed in this manner, these approaches did not have much practical application. More recent work has therefore focussed on reducing the amount of work that is required to show that protocols are composable. Meadows, in her analysis of the IKE protocol suite using the NRL Protocol Analyzer [179], proved that the different sub-protocols did not interact insecurely with each other by restricting attention to only those parts of the sub-protocols, which had a chance of subverting each other's security goals. Independently, Thayer, Herzog and Guttman used a similar insight to develop a technique for proving composition results using their strand space model [249]. Their technique consisted in showing that a set of terms generated by one protocol can never be accepted by principals executing the other protocol. The techniques used for choosing the set of terms, however, is specific to the protocols in [108]. A somewhat different approach is used by Lynch [170] to prove that the composition of a simple shared key communication protocol and the Diffie-Hellman key distribution protocol is secure. Her model uses I/O automata and the protocols are shown to compose if adversaries are only passive eavesdroppers.

A different approach to composition is taken in the Protocol Composition Logic (PCL) project discussed in an earlier section. PCL supports compositional reasoning about security protocols, including parallel composition of different protocols, and sequential composition of protocol steps. For example, many protocols assume that long-term cryptographic keys have been properly distributed to protocol agents. PCL allows proofs of key-distribution protocols to be combined with proofs for protocols that use these keys. Technically, protocols compose securely in parallel if they satisfy each other's invariants while sequential composition requires, in addition, that the postcondition of the first protocol implies the precondition of the second (see [88] for further details and examples). Another aspect of PCL is a composition method based on protocol templates [86], which are "abstract" protocols containing function variables for some of the operations used to construct messages. In the template method, correctness of a protocol template may be established under certain assumptions about these function variables. Then, a proof for an actual protocol is obtained by replacing the function variables with combinations of operations that satisfy the proof assumptions.

PCL appears to scale well to industrial protocols of five to twenty messages (or more), in part because PCL proofs appear to be relatively short (for formal proofs), and it has been successfully applied to a number of industry standards including SSL/TLS, IEEE 802.11i [128] and Kerberos V5 [221]. The PCL composition theorems were particularly useful in carrying out these larger-scale case studies. In recent work, Lowe and Auty take a related approach [25].

It is well known that many natural security properties (e.g., noninterference) are not preserved either under composition or under refinement. This has been extensively explored using trace-based modelling techniques [172; 173; 174; 175; 176], using properties that are not first-order predicates over traces, but second-order predicates over sets of traces that may not have closure properties corresponding to composition and refinement. In contrast, security properties expressed in PCL are safety properties over sets of traces that satisfy safety invariants, thus avoiding these negative results about composability.

There are some important differences between the PCL approach to compositional reasoning and alternative approaches such as universal composability [58; 210]. In universal composability, properties of a protocol are stated in a strong form using a simulation-style definition so that the property will be preserved under a wide class of composition operations. In contrast, PCL proofs proceed from various assumptions, including invariants that are assumed to hold in any environment in which the protocol operates. The ability to reason about protocol parts under assumptions about the way they will be used offers greater flexibility and appears essential for developing modular proofs about certain classes of protocols.

Finally, we note that although there are some similarities between the composition paradigm of PCL and the assume-guarantee paradigm in distributed computing [190], there is also one important difference. In PCL, while composing protocols, we check that each protocol respects the invariants of the other. This step involves an induction argument over the steps of the two protocols. There is no reasoning about attacker actions. One way to see the similarity with assume-guarantee is that each protocol is proved secure assuming some property of the other protocol and then discharging this assumption. The difference lies in the fact that the assumption made does not depend on the attacker although the environment for each protocol includes the attacker in addition to the other protocol.

5.4.2 COMPUTATIONAL SOUNDNESS

While there has been significant research on analysis of cryptographic protocols over the last three decades, the field was until recently divided into two largely disjoint communities based on the model and method used for analysis. The *symbolic model* [97; 201], favored by researchers working on logical methods in computer science, assumes perfect cryptography and a non-deterministic adversary who controls the network. The approaches described in the previous sections use this model. Security properties like *authentication* and *secrecy* are typically formulated as trace properties and a proof involves showing that every execution satisfies the desired property; the central proof technique is *induction*. The *computational model* [115] used in cryptographic studies, on the other hand, provides more detailed security definitions for primitives like encryption, signature, and message authentication codes [37] as well as for protocols for authentication and key exchange [36; 38; 60; 241]. One way of specifying security involves *games* between a challenger and an adversary: a primitive or protocol is secure if for all polynomial time adversaries, the probability that the adversary wins the game is very low (bounded from above by a negligible function of the security parameter, which determines the running time of all parties as well as the length of keys, random coins, etc.). Proofs proceed by *reduction*: assuming that there exists an adversary that defeats the protocol security game with non-negligible probability, a different adversary is constructed that defeats the security game for at least one of the primitives used in the protocol with non-negligible probability. Proofs in the computational model provide finer-grained and more realistic security guarantees than proofs in the symbolic model, but they are significantly more difficult to produce and verify.

Several recent projects have aimed to bridge the gap between these two lines of work. The overarching goal is to use high-level symbolic analysis techniques that are amenable to mechanization while providing the more realistic security guarantees afforded by the computational model. This research program on *computational soundness of symbolic methods* began with an equational logic of encryption developed by Abadi and Rogaway [12]. While this and related results [15; 33; 183] were restricted to passive adversaries, subsequent work has developed computationally sound analysis methods for network protocols interacting with an active, probabilistic polynomial time adversary. Backes *et al* [27; 28] use a simulation-based model to show that if the cryptographic primitives are sufficiently strong, then the "ideal" (symbolic) world is indistinguishable from the "real" (computational) world to any polynomially bounded environment. The "symbolic" model used in these papers involves extending the standard symbolic model with certain computational artifacts such as length of ciphertexts. A related approach is taken by Canetti *et al* [59]. Micciancio and Warinschi [184] establish a correspondence theorem between the standard symbolic and computational models using game-based definitions for computational security. The proof proceeds by reduction: a protocol attack in the computational model is translated into an attack on the underlying encryption scheme with non-negligible probability, assuming that the attack is not symbolic. There are a number of other results along these lines [76; 131; 139]. Laud [157] presents a protocol language and a type system such that if a protocol type checks then it preserves the computational secrecy of messages entered by the users. The semantics of the language is based on the work of Backes *et al* [28].

Datta *et al* [90; 91; 219; 220] have developed Computational PCL, a logic that supports direct reasoning about the computational model of cryptographic protocols. This work builds on previous work on PCL (described in Section 5.2). Security properties are specified using predicates whose semantics is defined with respect to the computational model. Proofs are carried out using a proof system that codifies high-level reasoning principles including composition theorems. The soundness of the proof system is established using reduction; parts of this proof are similar to the proof by Micciancio and Warinschi [184]. One feature of this approach is that it allows proofs of protocol properties from weaker assumptions about the cryptographic primitives. For example, it is possible to prove security for secure sessions assuming IND-CPA rather than IND-CCA security of encryption [91]. Another feature is that it is possible to express properties of primitives such as Diffie-Hellman and XOR for which correspondence theorems are either not known or impossible [26; 29; 220]. In a related project, Impagliazzo and Kapron present a logic for reasoning about cryptographic constructions [138]. Their focus is on primitives such as pseudorandom functions and not on network protocols.

A language-based approach to computational protocol analysis is taken by Mitchell, Scedrov and co-authors [165; 192; 193; 212]. They develop a process calculus for expressing probabilistic polynomial time protocols, a specification method based on a compositional form of equivalence, and an equational reasoning system for establishing equivalence between processes. In subsequent work, Blanchet *et al* have developed and implemented additional proof techniques, based on sequences of games, in a tool called CryptoVerif [40; 41].

Computationally sound protocol analysis is an active research area with papers being regularly published at the time of writing this section. While we have covered a representative set of results in this section, this is by no means a comprehensive survey of this area.

5.5   CONCLUSIONS

The research area of security protocol analysis has come of age. Building on work over the last three decades, we now have theoretically well-founded models and analysis techniques and tools. In the last five to ten years, these methods have been successfully applied to industrial protocols, in several cases identifying serious vulnerabilities and having an impact on industry standards. There has been significant advances in algorithmic analysis techniques as well as in logical proof methods. We have made progress on modular reasoning about security protocols although several challenges remain in this space, in particular, in mechanizing compositional reasoning. The results on computational soundness represent another major sub-area of current interest that involves analysis techniques from two largely disjoint communities—logic and cryptography—and improves further the fidelity of formal protocol analysis. The trend towards networked computing environments and the importance of secure communication continues to grow with the internet, and wireless, mobile, and sensor networks. We, therefore, believe that security protocol analysis will remain a lively research area for years to come, providing problems that are both intellectually challenging and have practical import.

APPENDIX A

Formalizing Static Analysis

A.1 PROGRAMS

A program is often specified by a *control flow graph (CFG)*—i.e., a directed graph $G = (V, E)$, where V is a set of program locations, and $E \subseteq V \times V$ is a set of edges that represent the flow of control. A node $n_e \in V$ denotes a unique entry point of the program. Node n_e is not allowed to have predecessors.

The state of a program is often modeled using a fixed, finite set of variables, *Vars*, whose values range over a set $\mathbb{V}$. We assume that the set $\mathbb{V}$ is application specific, e.g., $\mathbb{V}$ might be the set of integers ($\mathbb{Z}$), the set of rationals ($\mathbb{Q}$), the set of reals ($\mathbb{R}$), or the set of 32-bit two's-complement machine integers (int). We will use $\mathbb{B}$ to denote the set of Boolean values, i.e., $\mathbb{B} = \{true, false\}$.

A *program state S* is a function that maps each program variable to a corresponding value, i.e., $S : Vars \to \mathbb{V}$. We will use $\Sigma = Vars \to \mathbb{V}$ to denote the set of all possible program states.

A.1.1 EXPRESSIONS AND CONDITIONALS

We do not impose any *a priori* restrictions on the expressions and conditional expressions that can be used in the program: that is $x + y$, $\sqrt[3]{y}$, $x \bmod 5 = 0$, and $\sin^2(x) + \cos^2(x) = 1$ all represent valid numeric expressions or conditional expressions. Their semantics is defined by a family of functions $[\![\cdot]\!]$, which map the values assigned to the free variables in the expression to the result of the expression. That is, $[\![x + y]\!](3, 5)$ yields 8 and $[\![x \bmod 5 = 0]\!](25)$ yields *true*.[1] More formally, let Φ denote the set of all valid expressions, and let $\phi(v_1, \ldots, v_k) \in \Phi$ be a k-ary expression. Then, the semantics of ϕ is given by a function

$$[\![\phi(v_1, \ldots, v_k)]\!] : \mathbb{V}^k \to \mathbb{V}.$$

Similarly, let Ψ denote the set of all valid conditional expressions, and let $\psi(v_1, \ldots, v_k) \in \Psi$ be a k-ary conditional expression. Then the semantics of ψ is given by a function

$$[\![\psi(v_1, \ldots, v_k)]\!] : \mathbb{V}^k \to \mathbb{B}.$$

A.1.2 SUPPORT FOR NONDETERMINISM

Sometimes, it is convenient to be able to specify a certain degree of non-determinism in the program. For instance, nondeterminism can be used to model the effects of the environment, e.g., to model

---

[1] By convention, in our examples values are bound to variables in alphabetical order.

user input. We support nondeterminism by allowing a special constant "?" to be used within an expression: "?" chooses a value from $\mathbb{V}$ nondeterministically. To accommodate "?", we lift the semantics of expressions and conditional expressions to return sets of values: e.g., $[\![x+?]\!]_{ND}$ yields $\mathbb{V}$ for any value of x, and $[\![? \textbf{ mod } 3]\!]_{ND}$ yields the set $\{0, 1, 2\}$.

Without loss of generality, let expression $\phi(v_1, \ldots, v_k)$ have r occurrences of "?" in it. Let $\tilde{\phi}(v_1, \ldots, v_k, w_1, \ldots, w_r)$ denote an expression obtained by substituting each occurrence of "?" in ϕ with a fresh variable $w_i \notin Vars$. Then, the semantics for ϕ is defined as follows (let $\bar{\alpha} \in \mathbb{V}^k$):

$$[\![\phi(v_1, \ldots, v_k)]\!]_{ND}(\bar{\alpha}) = \left\{ [\![\tilde{\phi}(v_1, \ldots, v_k, w_1, \ldots, w_r)]\!](\bar{\alpha}, \bar{\beta}) \mid \bar{\beta} \in \mathbb{V}^r \right\}$$

The nondeterministic semantics for conditional expressions is defined similarly.

A.1.3   EVALUATION OF EXPRESSIONS AND CONDITIONAL EXPRESSIONS

We define the semantics for evaluating expressions and conditional expressions in a program state in a straightforward way. Let $S \in \Sigma$ denote an arbitrary program state, and let $\phi(v_1, \ldots, v_k) \in \Phi$, where $v_i \in Vars$, be an expression. The function $[\![\phi(v_1, \ldots, v_k)]\!] : \Sigma \to \wp(\mathbb{V})$ is defined as follows:

$$[\![\phi(v_1, \ldots, v_k)]\!](S) = [\![\phi(v_1, \ldots, v_k)]\!]_{ND}(S(v_1), \ldots, S(v_k)).$$

Similarly, let $\psi(v_1, \ldots, v_k) \in \Psi$, where $v_i \in Vars$, be a conditional. The function $[\![\psi(v_1, \ldots, v_k)]\!] : \Sigma \to \wp(\mathbb{B})$ is defined as follows:

$$[\![\psi(v_1, \ldots, v_k)]\!](S) = [\![\psi(v_1, \ldots, v_k)]\!]_{ND}(S(v_1), \ldots, S(v_k)).$$

From now on, we will omit the lists of free variables when referring to expressions and conditional expressions, unless those lists are important to the discussion.

A.1.4   CONCRETE SEMANTICS OF A PROGRAM

The function $\Pi_G : E \to (\Sigma \to \Sigma)$ assigns to each edge in the CFG the concrete semantics of the corresponding program-state transition. Two types of transitions are allowed:

- **Assignment transition, $\bar{x} := \bar{\phi}$:** An assignment transition allows multiple variables to be updated in parallel; i.e., $\bar{x} \in Vars^m$, where $1 \leq m \leq |Vars|$, with an additional constraint that each variable may appear at most once in $\bar{x}$. Also, $\bar{\phi} \in \Psi^m$. As an example, the assignment transition $\langle x, y \rangle := \langle y, x + 1 \rangle$ assigns the value of variable y to variable x, and the value of $x + 1$ to variable y. The semantics of an assignment transition is defined as follows (we use $x[i]$ to denote the i-th component of vector $\bar{x}$):

$$[\![\bar{x} := \bar{\phi}]\!](S) = \left\{ S' \in \Sigma \mid \forall v \in Vars \left[\begin{array}{ll} S'(v) \in [\![\phi[i]]\!]_{ND}(S) & \text{if } v = x[i], \ i \in [1, m] \\ S'(v) = S(v) & \text{otherwise} \end{array} \right] \right\}$$

Typically, only a single variable is updated by an assignment transition. In that case, we will omit the vector notation, e.g., $x := x + 1$. The assignment transition $x := ?$ "forgets" the value of variable x.

- **Assume transition, *assume*(ψ):** An assume transition filters out program states in which the condition $\psi \in \Psi$ does not hold. For uniformity, we define the semantics of the transition to map a program state to a singleton set containing that program state, if the condition ψ holds in that state; otherwise, a program state is mapped to the empty set

$$\llbracket assume(\psi) \rrbracket(S) = \begin{cases} \{S\} & \text{if } true \in \llbracket \psi \rrbracket_{ND}(S) \\ \emptyset & \text{otherwise} \end{cases}$$

The semantics of program-state transitions is extended trivially to operate on sets of program states

$$\Pi_G(e)(SS) = \bigcup_{S \in SS} \Pi_G(e)(S),$$

where $e \in E$ and $SS \subseteq \Sigma$.

A.1.5 THE CONCRETE COLLECTING SEMANTICS OF A PROGRAM

We will use maps $\Theta : V \to \wp(\Sigma)$ from program locations to program states to collect the sets of reachable states. Let $\Theta_{\triangleright}$ denote a map that represents the initial state of the program. Typically, we assume that the program execution starts at the entry point n_e, and that any state is possible at n_e:

$$\Theta_{\triangleright}(v) = \begin{cases} \Sigma & \text{if } v = n_e \\ \emptyset & \text{otherwise} \end{cases}, \quad \text{for all } v \in V$$

The *collecting semantics* of a program (that is, a function that maps each program location to the set of program states that arise at that location), is given by the least map $\Theta_{\star}$ that satisfies the following conditions:

$$\Theta_{\star}(v) \supseteq \Theta_{\triangleright}(v), \quad \text{and} \quad \Theta_{\star}(v) = \bigcup_{\langle u,v \rangle \in E} \Pi_G(\langle u, v \rangle)(\Theta_{\star}(u)), \text{ for all } v \in V$$

The goal of program analysis is, given $\Theta_{\triangleright}$, compute $\Theta_{\star}$. However, this problem is generally undecidable.

A.2 ABSTRACTION AND ABSTRACT DOMAINS

Program analysis sidesteps undecidability by using abstraction: sets of program states are over-approximated by elements of an abstract domain $\mathbb{D} = \langle D, \sqsubseteq, \top, \bot, \sqcup, \sqcap \rangle$, where $\sqsubseteq$ is a binary relation that is reflexive, transitive, and anti-symmetric: it imposes a partial order on D; $\top$ and $\bot$ denote, respectively, the greatest and the least elements of D with respect to $\sqsubseteq$; $\sqcup$ and $\sqcap$ denote the least upper bound (*join*) operator and the greatest lower bound (*meet*) operator, respectively.

The elements of $\mathbb{D}$ are linked to sets of concrete program states by a pair of functions $\langle \alpha, \gamma \rangle$, where $\alpha : \wp(\Sigma) \to D$ is an *abstraction function*, which constructs an approximation for a set of states, and $\gamma : D \to \wp(\Sigma)$ is a *concretization function*, which gives meaning to domain elements; The functions α and γ are chosen to form a Galois connection, that is

$$\forall S \in \wp(\Sigma) \; \forall d \in D \; \left[\; \alpha(S) \sqsubseteq d \; \Leftrightarrow \; S \subseteq \gamma(d) \; \right]$$

It follows immediately that $d_1 \sqsubseteq d_2 \Rightarrow \gamma(d_1) \subseteq \gamma(d_2)$; thus, the smaller the abstract-domain element with respect to $\sqsubseteq$, the smaller the set of states that it denotes. The least element, $\bot$, typically represents the empty set, i.e., $\gamma(\bot) = \emptyset$. The greatest element, $\top$, represents the entire set of program states, i.e., $\gamma(\top) = \Sigma$. The join operator and the meet operator soundly approximate set union and set intersection, respectively.

A.2.1   ABSTRACT SEMANTICS OF A PROGRAM

To perform program analysis, the program-state transitions that are associated with the edges of a control flow graph also need to be abstracted. We will use the map $\Pi_G^{\sharp} : E \to (D \to D)$ to specify corresponding abstract transformers for each edge in the CFG. We say that $\Pi_G^{\sharp}$ is a *sound approximation* of Π_G if the following condition holds:

$$\forall e \in E \; \forall d \in D \; \left[\; \Pi_G(e)(\gamma(d)) \; \subseteq \; \gamma(\Pi_G^{\sharp}(e)(d)) \; \right].$$

A.2.2   ABSTRACT COLLECTING SEMANTICS

To refer to abstract states at multiple program locations, we define *abstract-state maps* $\Theta^{\sharp} : V \to D$. We also define the operations α, γ, $\sqsubseteq$, and $\sqcup$ for $\Theta^{\sharp}$ as point-wise extensions of the corresponding operations for the abstract domain $\mathbb{D}$.

Program analysis computes a sound approximation for the set of program states that are reachable from $\Theta_{\triangleright}$. Typically, the result of program analysis is an abstract-state map $\Theta_{\star}^{\sharp}$ that satisfies the following property

$$\forall v \in V : \; \left[\Theta_{\triangleright}^{\sharp}(v) \sqcup \bigsqcup_{\langle u, v \rangle \in E} \Pi_G^{\sharp}(\langle u, v \rangle)(\Theta_{\star}^{\sharp}(u)) \right] \sqsubseteq \Theta_{\star}^{\sharp}(v), \tag{A.1}$$

where $\Theta_{\triangleright}^{\sharp} = \alpha(\Theta_{\triangleright})$ is the approximation for the set of initial states of the program. It follows trivially from the definition that the resulting approximation is sound, that is $\Theta_{\star}(v) \subseteq \gamma(\Theta_{\star}^{\sharp}(v))$ for all $v \in V$.

A.3   ITERATIVE COMPUTATION

This section explains one methodology for computing $\Theta_{\star}^{\sharp}$, the abstract collecting semantics of a program, by means of an iterative process of successive approximation.

A.3.1   KLEENE ITERATION

If the abstract domain $\mathbb{D}$ and the set of abstract transformers in $\Pi_G^\sharp$ possess certain algebraic properties, then $\Theta_\star^\sharp$ can be obtained by computing the following sequence of abstract-state maps until it stabilizes:

$$\Theta_0^\sharp = \Theta_\rhd^\sharp \quad \text{and} \quad \Theta_{i+1}^\sharp(v) = \bigsqcup_{\langle u,v \rangle \in E} \Pi_G^\sharp(\langle u,v \rangle)(\Theta_i^\sharp(u)) \qquad \text{(A.2)}$$

In particular, the abstract transformers in $\Pi_G^\sharp$ must be *monotone*, i.e.,

$$\forall e \in E \ \forall d_1, d_2 \in D \ \left[\ d_1 \sqsubseteq d_2 \ \Rightarrow \ \Pi_G^\sharp(e)(d_1) \sqsubseteq \Pi_G^\sharp(e)(d_2) \ \right].$$

Also, to ensure termination, the abstract domain $\mathbb{D}$ must satisfy the *ascending-chain condition*; i.e., every sequence of elements $(d_k) \in D$ such that $d_1 \sqsubseteq d_2 \sqsubseteq d_3 \sqsubseteq \ldots$ must eventually become stationary.

Additionally, if the transformers in $\Pi_G^\sharp$ *distribute* over the join operator of $\mathbb{D}$, i.e., if

$$\forall e \in E \ \forall d_1, d_2 \in D \ \left[\ \Pi_G^\sharp(e)(d_1 \sqcup d_2) \ = \ \Pi_G^\sharp(e)(d_1) \sqcup \Pi_G^\sharp(e)(d_2) \ \right],$$

then the solution obtained for $\Theta_\star^\sharp$ is the most precise (i.e., least) solution that satisfies Eqn. (A.1).

A.3.2   WIDENING

If the domain does not satisfy the ascending-chain condition, then the sequence defined in Eqn. (A.2) may not necessarily converge. To make use of such domains in practice, an extrapolation operator, called *widening*, is defined. A widening operator (∇_k) must possess the following properties:

- For all $d_1, d_2 \in D$, for all $i \geq 0, d_1, d_2 \sqsubseteq d_1 \nabla_i d_2$.

- For any ascending chain $(d_k) \in D$, the chain defined by $d'_0 = d_0$ and $d'_{i+1} = d'_i \nabla_i d_{i+1}$ is not strictly increasing.

To make an iterative computation converge, a set $W \subseteq V$ of widening points is identified: W must be chosen in such a way that every loop in Π is *cut* by a node in W. Typically, W is chosen to contain the heads of all of the loops in the program. The iteration proceeds as follows:

$$\Theta_0^\sharp = \Theta_\rhd^\sharp \quad \text{and} \quad \Theta_{i+1}^\sharp(v) = \Theta_i^\sharp(v) \bowtie \bigsqcup_{\langle u,v \rangle \in E} \Pi_G^\sharp(\langle u,v \rangle)(\Theta_i^\sharp(u)) \qquad \text{(A.3)}$$

where $\bowtie$ is ∇_i if $v \in W$ and $\sqcup$, otherwise.

Intuitively, widening attempts to guess loop invariants by observing program states that arise on the first few iterations of the loop. Typically, delaying the application of widening for a few iterations tends to increase the precision of the analysis. To delay the application of widening for k

iterations, the widening operator can be redefined as follows:

$$\nabla_i^{delay[k]} = \begin{cases} \sqcup & \text{if } i < k \\ \nabla_{i-k} & \text{otherwise} \end{cases}$$

(Often the definition of the widening operator is independent from the iteration number on which the operator is invoked. In that case, one typically denotes the widening operator by ∇, with no subscripts.)

A.3.3   NARROWING

The limit of the sequence in Eqn. (A.3), which we will denote by $\Theta_\triangleleft^\sharp$, is sound, but generally overly conservative. It can be refined to a more precise solution by computing a *descending-iteration sequence*:

$$\Theta_0^\sharp = \Theta_\triangleleft^\sharp \quad \text{and} \quad \Theta_{i+1}^\sharp(v) = \bigsqcup_{\langle u, v \rangle \in E} \Pi_G^\sharp(\langle u, v \rangle)(\Theta_i^\sharp(u)) \tag{A.4}$$

However, for this sequence to converge, the abstract domain must satisfy the *descending-chain condition*, that is, the domain must contain no infinite strictly-decreasing chains. If the abstract domain does not satisfy this property, convergence may be enforced with the use of a *narrowing operator*. A narrowing operator, (Δ_k) must possess the following properties:

- For all $d_1, d_2 \in D$, for all $i \geq 0$, $d_2 \sqsubseteq d_1 \Rightarrow [\, d_2 \sqsubseteq d_1 \, \Delta_i \, d_2 \, \wedge \, d_1 \, \Delta_i \, d_2 \sqsubseteq d_1 \,]$.

- For any descending chain $(d_k) \in D$, the chain defined by $d_0' = d_0$ and $d_{i+1}' = d_i' \, \Delta_i \, d_{i+1}$ is not strictly decreasing.

The computation of the descending sequence proceeds as follows:

$$\Theta_0^\sharp = \Theta_\triangleleft^\sharp \quad \text{and} \quad \Theta_{i+1}^\sharp(v) = \Theta_i^\sharp(v) \bowtie \bigsqcup_{\langle u, v \rangle \in E} \Pi_G^\sharp(\langle u, v \rangle)(\Theta_i^\sharp(u)) \tag{A.5}$$

where $\bowtie$ is Δ_i if $v \in W$ and $\sqcap$, otherwise.

The overall result of the analysis, $\Theta_\star^\sharp$, is the limit of the sequence computed in accordance with Eqn. (A.5). Typically, $\Theta_\star^\sharp$ is not the most precise abstraction of the program's collecting semantics with respect to the property in Eqn. (A.1).

Meaningful narrowing operators are much harder to define than widening operators; thus, many abstract domains do not provide them. For those domains, the descending-iteration sequence from Eqn. (A.4) is, typically, truncated after some fixed number of iterations. One way to truncate the iteration sequence is to define a simple domain-independent narrowing operator that cuts off the decreasing sequence after some number of iterations; i.e., to truncate after k iterations, the narrowing operator can be defined as follows:

$$d_1 \, \Delta_i \, d_2 = \begin{cases} d_2 & \text{if } i < k \\ d_1 & \text{otherwise} \end{cases}$$

A.3.4   CHAOTIC ITERATION

Computing the sequence of abstract-state maps according to Eqns. (A.3) and (A.5) is not efficient in practice: on each step, the mappings for *all* program points are recomputed even though only a small number of them actually change value. *Chaotic iteration* allows one to speed up the computation by only updating a value at single program point on each iteration of the analysis: given a *fair* sequence of program points $\sigma \in V^{\infty}$ (that is, a sequence in which each program point appears infinitely often), the ascending-iteration sequence can be computed as follows:

$$\Theta^{\sharp}_{i+1}(v) = \begin{cases} \Theta^{\sharp}_i(v) \; \nabla_i \; \bigsqcup_{\langle u,v\rangle \in E} \Pi^{\sharp}_G(\langle u,v\rangle)(\Theta^{\sharp}_i(u)) & \text{if } v = \sigma[i] \text{ and } v \in W \\ \Theta^{\sharp}_i(v) \; \sqcup \; \bigsqcup_{\langle u,v\rangle \in E} \Pi^{\sharp}_G(\langle u,v\rangle)(\Theta^{\sharp}_i(u)) & \text{if } v = \sigma[i] \text{ and } v \notin W \\ \Theta^{\sharp}_i(v) & \text{otherwise } (v \neq \sigma[i]) \end{cases} \qquad (A.6)$$

The descending iteration sequence is computed in a similar way, with ∇ replaced with Δ, and $\sqcup$ replaced with $\sqcap$.

The use of chaotic iteration raises the question of an effective *iteration strategy*—that is, an order in which program points are visited that minimizes the amount of work that the analysis performs. This problem was addressed by Bourdoncle [49]. He proposed a number of successful iteration strategies based on the idea of using a *weak topological order (WTO)* of the nodes in the program's control-flow graph. A WTO is a hierarchical structure that decomposes *strongly-connected components (SCCs)* in the graph into a set of nested WTO components. (In structured programs, WTO components correspond to the loops in the program.) Of particular interest is the *recursive iteration strategy*.

Definition A.3.1 (Recursive Iteration Strategy [49]) *The recursive iteration strategy recursively stabilizes the sub-components of a WTO component before stabilizing that WTO component.* □

The recursive iteration strategy forces an analyzer to stay within a loop until the stable solution for that loop is obtained; only then is the analyzer allowed to return to the outer loop. The recursive iteration strategy has the nice property that the analyzer only needs to check for stabilization at the head of the corresponding WTO component [49] (roughly, only at the head of each loop).

Bibliography

[1] http://verify.stanford.edu/dill/murphi.html. 118

[2] http://www.cs.utexas.edu/~shmat/murphi/ssl.m 118

[3] bugtraq. www.securityfocus.com 52

[4] CERT/CC advisories. www.cert.org/advisories 52

[5] The twenty most critical internet security vulnerabilities. www.sans.org/top20. 35

[6] A guide to understanding discretionary access control in trusted systems, Sept. 1987. National Computer Security Center Report NCSC-TG-003. 69

[7] Aleph-one. smashing the stack for fun and profit. Nov 1996. Phrack Magazine. 35

[8] IEEE Standard 802.11-1999. Local and metropolitan area networks - specific requirements - part 11: Wireless LAN Medium Access Control and Physical Layer specifications., 2004. 2, 3, 99

[9] M. Abadi, B. Blanchet, and C. Fournet. Just Fast Keying in the Pi Calculus. In *Programming Languages and Systems: Proceedings of the 13th European Symposium on Programming (ESOP'04)*, volume 2986 of *Lecture Notes on Computer Science*, pages 340–354. Springer Verlag, 2004. DOI: 10.1145/1266977.1266978 119

[10] M. Abadi and C. Fournet. Mobile values, new names, and secure communication. In *28th ACM Symposium on Principles of Programming Languages*, pages 104–115, 2001. DOI: 10.1145/360204.360213 99, 101, 102, 119

[11] M. Abadi and A. Gordon. A calculus for cryptographic protocols: the spi calculus. *Information and Computation*, 148(1):1–70, 1999. Expanded version available as SRC Research Report 149 (January 1998). DOI: 10.1006/inco.1998.2740 99, 101, 102, 119

[12] M. Abadi and P. Rogaway. Reconciling two views of cryptography (the computational soundness of formal encryption). *Journal of Cryptology*, 15(2):103–127, 2002. 4, 122

[13] S. Abiteboul, R. Hull, and V. Vianu. *Foundations of Databases*. Addison-Wesley, 1995. 33

[14] S. Abramsky and C. Hankin. *Abstract Interpretation of Declarative Languages*. John Wiley and Sons, 1987. 6

[15] P. Adão, G. Bana, and A. Scedrov. Computational and information-theoretic soundness and completeness of formal encryption. In *Proc. of the 18th IEEE Computer Security Foundnations Workshop*, pages 170–184, 2005. DOI: 10.1109/CSFW.2005.13 4, 122

[16] A. Aho, R. Sethi, and J. Ullman. *Compilers: Principles, Techniques and Tools*. Addison-Wesley, 1985. 6

[17] R. K. Ahuja, T. L. Magnanti, and J. B. Orlin. *Network Flows: Theory, Algorithms and Applications*. Prentice Hall, 1993. 43

[18] R. M. Amadio and D. Lugiez. On the reachability problem in cryptographic protocols. In *CONCUR*, pages 380–394, 2000. 118

[19] P. Ammann and R. S. Sandhu. Safety analysis for the extended schematic protection model. In *Proceedings of the 1991 IEEE Symposium on Security and Privacy*, pages 87–97, May 1991. DOI: 10.1109/RISP.1991.130777 63, 69

[20] P. Ammann and R. S. Sandhu. The extended schematic protection model. *Journal of Computer Security*, 1(3-4):335–383, 1992. 63, 69

[21] L. O. Andersen. *Program Analysis and Specialization for the C Programming Language*. PhD thesis, DIKU, Univ. of Copenhagen, May 1994. 37, 53

[22] E. D. Anderson and K. D. Anderson. Presolving in linear programming. *Mathematical Programming*, 71(2):221–245, 1995. DOI: 10.1007/BF01586000 46

[23] ANSI. American national standard for information technology – role based access control. ANSI INCITS 359-2004, Feb. 2004. 69

[24] A. Armando, D. A. Basin, Y. Boichut, Y. Chevalier, L. Compagna, J. Cuéllar, P. H. Drielsma, P.-C. Héam, O. Kouchnarenko, J. Mantovani, S. Mödersheim, D. von Oheimb, M. Rusinowitch, J. Santiago, M. Turuani, L. Viganò, and L. Vigneron. The AVISPA tool for the automated validation of internet security protocols and applications. In *Computer Aided Verification, 17th International Conference, CAV 2005, Proceedings*, volume 3576 of *Lecture Notes in Computer Science*, pages 281–285. Springer, 2005. DOI: 10.1007/11513988_27 118

[25] M. Auty and G. Lowe. A calculus for security protocol development. Submitted for publication, 2006. 121

[26] M. Backes and B. Pfitzmann. Limits of the cryptographic realization of XOR. In *Proc. of the 10th European Symposium on Research in Computer Security*. Springer-Verlag, 2005. 123

[27] M. Backes and B. Pfitzmann. Relating symbolic and cryptographic secrecy. In *Proc. IEEE Symposium on Security and Privacy*, pages 171–182. IEEE, 2005. DOI: 10.1109/TDSC.2005.25 122

[28] M. Backes, B. Pfitzmann, and M. Waidner. A universally composable cryptographic library. Cryptology ePrint Archive, Report 2003/015, 2003. 122

[29] M. Backes, B. Pfitzmann, and M. Waidner. Limits of the reactive simulatability/uc of dolev-yao models with hashes. In *Proc. of the 11th European Symposium on Research in Computer Security*. Springer-Verlag, 2006. 123

[30] T. Ball and S. Rajamani. Bebop: A symbolic model checker for Boolean programs. In *Spin Workshop*, volume 1885 of *Lec. Notes in Comp. Sci.*, pages 113–130, 2000. 12

[31] T. Ball and S. Rajamani. Bebop: A path-sensitive interprocedural dataflow engine. In *Prog. Analysis for Softw. Tools and Eng.*, pages 97–103, June 2001. DOI: 10.1145/379605.379690 12

[32] D. A. Basin, S. Mödersheim, and L. Viganò. Ofmc: A symbolic model checker for security protocols. *Int. J. Inf. Sec.*, 4(3):181–208, 2005. DOI: 10.1007/s10207-004-0055-7 118

[33] M. Baudet, V. Cortier, and S. Kremer. Computationally Sound Implementations of Equational Theories against Passive Adversaries. In *Proceedings of the 32nd International Colloquium on Automata, Languages and Programming (ICALP'05)*, volume 3580 of *Lecture Notes in Computer Science*, pages 652–663. Springer, 2005. DOI: 10.1007/11523468_53 4, 122

[34] G. Bella, F. Massacci, L. C. Paulson, and P. Tramontano. Formal verification of cardholder registration in set. In *ESORICS*, pages 159–174, 2000. DOI: 10.1007/10722599_10 119

[35] G. Bella and L. C. Paulson. Kerberos version IV: Inductive analysis of the secrecy goals. In J.-J. Quisquater, editor, *Proceedings of the 5th European Symposium on Research in Computer Security*, pages 361–375, Louvain-la-Neuve, Belgium, 1998. Springer-Verlag LNCS 1485. 119

[36] M. Bellare, R. Canetti, and H. Krawczyk. A modular approach to the design and analysis of authentication and key exchange protocols. In *Proceedings of 30th Annual Symposium on the Theory of Computing*. ACM, 1998. DOI: 10.1145/276698.276854 122

[37] M. Bellare and P. Rogaway. Introduction to modern cryptography. In *Unpublished Lecture Notes*. http://www-cse.ucsd.edu/users/mihir/cse207/classnotes.html 122

[38] M. Bellare and P. Rogaway. Entity authentication and key distribution. In *Advances in Cryprtology - Crypto '93 Proceedings*, pages 232–249. Springer-Verlag, 1994. DOI: 10.1007/3-540-48329-2_21 103, 114, 122

[39] B. Blanchet. An Efficient Cryptographic Protocol Verifier Based on Prolog Rules. In *14th IEEE Computer Security Foundations Workshop (CSFW-14)*, pages 82–96. IEEE Computer Society, 2001. DOI: 10.1109/CSFW.2001.930138 119

[40] B. Blanchet. A computationally sound mechanized prover for security protocols. In *IEEE Symposium on Security and Privacy*, pages 140–154, 2006. DOI: 10.1109/TDSC.2007.1005 123

[41] B. Blanchet and D. Pointcheval. Automated security proofs with sequences of games. In *CRYPTO*, pages 537–554, 2006. DOI: 10.1007/11818175_32 123

[42] M. Blaze, J. Feigenbaum, J. Ioannidis, and A. Keromytis. The KeyNote trust-management system version 2. RFC 2704, Network Working Group, 1999. 76

[43] M. Blaze, J. Feigenbaum, J. Ioannidis, and A. D. Keromytis. The role of trust management in distributed systems. In *Secure Internet Programming*, volume 1603 of *Lecture Notes in Computer Science*, pages 185–210. Springer, 1999. DOI: 10.1007/3-540-48749-2_8 2

[44] M. Blaze, J. Feigenbaum, and J. Lacy. Decentralized trust management. In *Proc. IEEE Symp. on Sec. and Privacy*, Oakland, CA, 1996. DOI: 10.1109/SECPRI.1996.502679 76

[45] R. Bodik, R. Gupta, and V. Sarkar. ABCD: Eliminating array-bounds checks on demand. In *ACM SIGPLAN Conference on Programming Language Design and Implementation (PLDI)*, 2000. DOI: 10.1145/349299.349342 35, 61

[46] M. Boreale and M. G. Buscemi. A framework for the analysis of security protocols. In *CONCUR*, pages 483–498, 2002. DOI: 10.1007/3-540-45694-5_32 118

[47] A. Bouajjani, J. Esparza, and O. Maler. Reachability analysis of pushdown automata: Application to model checking. In *Proc. CONCUR*, volume 1243 of *Lec. Notes in Comp. Sci.*, pages 135–150. Springer-Verlag, 1997. DOI: 10.1007/3-540-63141-0_10 12, 17

[48] A. Bouajjani, J. Esparza, and T. Touili. A generic approach to the static analysis of concurrent programs with procedures. In *Princ. of Prog. Lang.*, pages 62–73, 2003. DOI: 10.1145/604131.604137 12, 19

[49] F. Bourdoncle. Efficient chaotic iteration strategies with widenings. In *Int. Conf. on Formal Methods in Prog. and their Appl.*, Lec. Notes in Comp. Sci. Springer-Verlag, 1993. DOI: 10.1007/BFb0039704 131

[50] C. Boyd and A. Mathuria. *Protocols for Authentication and Key Establishment*. Springer-Verlag, 2003. 105

[51] D. Brumley, J. Newsome, D. Song, H. Wang, and S. Jha. Towards automatic generation of vulnerability-based signatures. In *IEEE Symposium on Security and Privacy*, Oakland, California, May 2006. DOI: 10.1109/SP.2006.41 1

[52] R. Bryant. Graph-based algorithms for Boolean function manipulation. *IEEE Trans. on Comp.*, C-35(6):677–691, Aug. 1986. DOI: 10.1109/TC.1986.1676819 22

[53] J. Büchi. *Finite Automata, their Algebras and Grammars*. Springer-Verlag, 1988. D. Siefkes (ed.). 17

[54] T. Budd. Safety in grammatical protection systems. *International Journal of Computer and Information Sciences*, 12(6):413–430, 1983. DOI: 10.1007/BF00977968 63

[55] O. Burkart and B. Steffen. Model checking for context-free processes. In *Proc. CONCUR*, volume 630 of *Lec. Notes in Comp. Sci.*, pages 123–137. Springer-Verlag, 1992. DOI: 10.1007/BFb0084787 16

[56] M. Burrows, M. Abadi, and R. Needham. A logic of authentication. *ACM Transactions on Computer Systems*, 8(1):18–36, 1990. DOI: 10.1145/77648.77649 99, 102, 104

[57] C. Cadar, V. Ganesh, P. M. Pawlowski, D. L. Dill, and D. R. Engler. Exe: Automatically generating inputs of death. In *13th ACM Conference on Computer and Communications Security (CCS)*, 2006. DOI: 10.1145/1180405.1180445 1

[58] R. Canetti. Universally composable security: A new paradigm for cryptographic protocols. In *Proc. 42nd IEEE Symp. on the Foundations of Computer Science*. IEEE, 2001. Full version available at http://eprint.iacr.org/2000/067/. DOI: 10.1109/SFCS.2001.959888 121

[59] R. Canetti, L. Cheung, D. K. Kaynar, M. Liskov, N. A. Lynch, O. Pereira, and R. Segala. Time-bounded task-pioas: A framework for analyzing security protocols. In *DISC*, pages 238–253, 2006. 122

[60] R. Canetti and H. Krawczyk. Analysis of key-exchange protocols and their use for building secure channels. In *Proc. of EUROCRYPT 2001*, volume 2045 of *LNCS*, pages 453–474, 2001. 122

[61] I. Cervesato, A. Jaggard, A. Scedrov, J.-K. Tsay, and C. Walstad. Breaking and fixing public-key kerberos. In *Proceedings of the 11-th Annual Asian Computing Science Conference*, 2006.

DOI: 10.1016/j.ic.2007.05.005 3, 99

[62] H. Chen, N. Li, and Z. Mao. Analyzing and comparing the protection quality of security enhanced operating systems. In *Proceedings of the 2009 Network and Distributed System Security Symposium (NDSS'09)*, Feb. 2009. 2

[63] H. Chen and D. Wagner. Mops: an infrastructure for examining security properties of software. In *In Proceedings of the 9th ACM Conference on Computer and Communications Security (CCS)*, Washington, DC, November 2002. 1

[64] J. W. Chinnek and E. W. Dravinieks. Locating minimal infeasible constraint sets in linear programs. *ORSA Journal on Computing*, 3(2):157–168, 1991. DOI: 10.1287/ijoc.3.2.157 45

[65] T.-C. Chiueh and F.-H. Hsu. RAD: A compile-time solution to buffer overflow attacks. In *Proceedings of the 21st International Conference on Distributed Computing Systems (ICDCS)*, April 2001. DOI: 10.1109/ICDSC.2001.918971 35, 61

[66] M. Christodorescu, S. Seshia, S. Jha, D. Song, and R. E. Bryant. Semantics-aware malware detection. In *IEEE Symposium on Security and Privacy*, Oakland, California, May 2005. DOI: 10.1109/SP.2005.20 1

[67] Y.-H. Chu, J. Feigenbaum, B. LaMacchia, P. Resnick, and M. Strauss. REFEREE: Trust management for Web applications. *Computer Networks and ISDN Systems*, 29, 1997. DOI: 10.1016/S0169-7552(97)00009-3 76

[68] V. Chvátal. *Linear Programming*. W. H. Freeman and Co., New York, 2000. 42

[69] D. Clarke, J.-E. Elien, C. Ellison, M. Fredette, A. Morcos, and R. Rivest. Certificate chain discovery in SPKI/SDSI. *J. Comp. Sec.*, 2001. 77, 80, 83, 91, 92

[70] E. M. Clarke, S. Jha, and W. R. Marrero. Verifying security protocols with brutus. *ACM Trans. Softw. Eng. Methodol.*, 9(4):443–487, 2000. DOI: 10.1145/363516.363528 118

[71] E. Cohen. Taps: A first-order verifier for cryptographic protocols. In *CSFW*, pages 144–158, 2000. DOI: 10.1109/CSFW.2000.856933 119

[72] E. Cohen. First-order verification of cryptographic protocols. *Journal of Computer Security*, 11(2):189–216, 2003. 119

[73] J. Condit, M. Harren, S. McPeak, G. C. Necula, and W. Weimer. CCured in the Real World. In *ACM SIGPLAN Conference on Programming Language Design and Implementation (PLDI)*, June 2003. DOI: 10.1145/780822.781157 35, 61

[74] T. H. Cormen, C. E. Lieserson, R. L. Rivest, and C. Stein. *Introduction to Algorithms*. The MIT Press, Cambridge, MA, 2001. 48, 51

[75] V. Cortier, S. Delaune, and P. Lafourcade. A survey of algebraic properties used in cryptographic protocols. *Journal of Computer Security*, 14(1):1–43, 2006. 120

[76] V. Cortier and B. Warinschi. Computationally sound, automated proofs for security protocols. In *Proceedings of 14th European Symposium on Programming (ESOP'05)*, Lecture Notes in Computer Science, pages 157–171. Springer-Verlag, 2005. DOI: 10.1007/b107380 122

[77] P. Cousot and R. Cousot. Abstract interpretation: A unified lattice model for static analysis of programs by construction of approximation of fixed points. In *Princ. of Prog. Lang.*, pages

238–252, 1977. 5, 6, 7, 13, 29

[78] P. Cousot and R. Cousot. Static determination of dynamic properties of recursive procedures. In E. Neuhold, editor, *Formal Descriptions of Programming Concepts, (IFIP WG 2.2, St. Andrews, Canada, August 1977)*, pages 237–277. North-Holland, 1978. 11, 20

[79] P. Cousot and N. Halbwachs. Automatic discovery of linear constraints among variables of a program. In *Princ. of Prog. Lang.*, pages 84–96, 1978. DOI: 10.1145/512760.512770 29

[80] C. Cowan, S. Beaattie, R.-F. Day., C. Pu, P. Wagle, and E. Walthinsen. Automatic detection and prevention of buffer overflow attacks. In *Proceedings of the 7th USENIX Security Symposium*, 1998. 35, 61

[81] J. Crampton and G. Loizou. Administrative scope and role hierarchy operations. In *Proceedings of Seventh ACM Symposium on Access Control Models and Technologies (SACMAT 2002)*, pages 145–154, June 2002. DOI: 10.1145/507711.507736 69

[82] J. Crampton and G. Loizou. Administrative scope: A foundation for role-based administrative models. *ACM Transactions on Information and System Security*, 6(2):201–231, May 2003. DOI: 10.1145/762476.762478 69

[83] J. R. Crandall, Z. Su, S. F. Wu, and F. T. Chong. On deriving unknown vulnerabilities from zero-day polymorphic and metamorphic worm exploits. In *In the proceedings of the 12th ACM Conference on Computer and Communications Security (CCS 2005)*, Alexandria, Virginia, November 2005. DOI: 10.1145/1102120.1102152 1

[84] R. Cytron, J. Ferrante, B. K. Rosen, M. N. Wegman, and F. K. Zadeck. Efficiently computing static single assignment form and the control dependence graph. *ACM Transactions on Programming Languages and Systems (TOPLAS)*, 13(4):452–490, October 1991. DOI: 10.1145/115372.115320 60

[85] G. B. Dantzig. *Linear Programming and Extensions.* Princeton University Press, Princeton, N.J., 1963. 42

[86] A. Datta, A. Derek, J. C. Mitchell, and D. Pavlovic. Abstraction and refinement in protocol derivation. In *Proceedings of 17th IEEE Computer Security Foundations Workshop*, pages 30–45. IEEE, 2004. DOI: 10.1109/CSFW.2004.4 120

[87] A. Datta, A. Derek, J. C. Mitchell, and D. Pavlovic. A derivation system and compositional logic for security protocols. *Journal of Computer Security*, 13(3):423–482, 2005. 102

[88] A. Datta, A. Derek, J. C. Mitchell, and A. Roy. Protocol composition logic (pcl). *Electr. Notes Theor. Comput. Sci.*, 172:311–358, 2007. DOI: 10.1016/j.entcs.2007.02.012 3, 99, 100, 102, 109, 110, 120

[89] A. Datta, A. Derek, J. C. Mitchell, and A. Roy. Protocol composition logic (pcl). *Electr. Notes Theor. Comput. Sci.*, 172:311–358, 2007. DOI: 10.1016/j.entcs.2007.02.012 108

[90] A. Datta, A. Derek, J. C. Mitchell, V. Shmatikov, and M. Turuani. Probabilistic polynomial-time semantics for a protocol security logic. In *Proceedings of the 32nd International Colloquium on Automata, Languages and Programming (ICALP '05)*, Lecture Notes in Computer Science, pages 16–29. Springer-Verlag, 2005. DOI: 10.1007/11523468_2 4, 122

[91] A. Datta, A. Derek, J. C. Mitchell, and B. Warinschi. Computationally sound compositional logic for key exchange protocols. In *Proceedings of 19th IEEE Computer Security Foundations Workshop*, pages 321–334. IEEE, 2006. DOI: 10.1109/CSFW.2006.9 4, 122, 123

[92] J. DeTreville. Binder, a logic-based security language. In *Proceedings of the 2002 IEEE Symposium on Security and Privacy*, pages 105–113. IEEE Computer Society Press, May 2002. DOI: 10.1109/SECPRI.2002.1004365 2

[93] J. Detreville. Binder, a logic-based security language. In *Proc. IEEE Symp. on Sec. and Privacy*, Oakland, CA, 2002. DOI: 10.1109/SECPRI.2002.1004365 76

[94] T. Dierks and C. Allen. The tls protocol version 1.0, 1999. RFC 2246. 2, 99

[95] W. Diffie, P. C. V. Oorschot, and M. J. Wiener. Authentication and authenticated key exchanges. *Designs, Codes and Cryptography*, 2:107–125, 1992. DOI: 10.1007/BF00124891 114

[96] D. L. Dill. The mur*hi* verification system. In *CAV*, pages 390–393, 1996. 118

[97] D. Dolev and A. Yao. On the security of public-key protocols. *IEEE Transactions on Information Theory*, 2(29), 1983. DOI: 10.1109/SFCS.1981.32 3, 101, 108, 121

[98] N. Dor, M. Rodeh, and M. Sagiv. CSSV: Towards a realistic tool for statically detecting all buffer overflows in C. In *Proceedings of the ACM SIGPLAN Conference on Programming Language Design and Implementation (PLDI)*, June 2003. DOI: 10.1145/780822.781149 35, 37, 60, 61

[99] W. F. Dowling and J. H. Gallier. Linear-time algorithms for testing the satisfiability of propositional horn formulae. *Journal of Logic Programming*, 1(3):267–284, 1984. DOI: 10.1016/0743-1066(84)90014-1 32

[100] N. Durgin, J. C. Mitchell, and D. Pavlovic. A compositional logic for protocol correctness. In *Proceedings of 14th IEEE Computer Security Foundations Workshop*, pages 241–255. IEEE, 2001. 106

[101] N. Durgin, J. C. Mitchell, and D. Pavlovic. A compositional logic for proving security properties of protocols. *Journal of Computer Security*, 11:677–721, 2003. 102, 106

[102] N. A. Durgin, P. Lincoln, J. C. Mitchell, and A. Scedrov. Multiset rewriting and the complexity of bounded security protocols. *Journal of Computer Security*, 12(2):247–311, 2004. 101, 119

[103] C. Ellison. SPKI/SDSI certificates. Available at "http://world.std.com/ cme/html/spki.html". 2

[104] C. Ellison. SPKI requirements. RFC 2692, Network Working Group, 1999. 76, 77

[105] C. Ellison. Personal communication to S. Jha, T. Reps, and S. Schwoon, Apr. 2003. 91

[106] C. Ellison, B. Frantz, B. Lampson, R. Rivest, B. Thomas, and T. Ylonen. SPKI certificate theory. RFC 2693, Network Working Group, 1999. 2, 76, 77, 79, 84, 91

[107] J. Esparza, D. Hansel, P. Rossmanith, and S. Schwoon. Efficient algorithms for model checking pushdown systems. In *Computer Aided Verif.*, volume 1855 of *Lec. Notes in Comp. Sci.*, pages 232–247, July 2000. DOI: 10.1007/10722167_20 17

[108] F. J. T. Fábrega, J. C. Herzog, and J. D. Guttman. Strand spaces: Why is a security protocol correct? In *Proceedings of the 1998 IEEE Symposium on Security and Privacy*, pages 160–171, Oakland, CA, May 1998. IEEE Computer Society Press. 106, 119, 120

[109] A. Finkel, B.Willems, and P. Wolper. A direct symbolic approach to model checking pushdown systems. *Elec. Notes in Theor. Comp. Sci.*, 9, 1997. DOI: 10.1016/S1571-0661(05)80426-8 12, 17

[110] R. Focardi, R. Gorrieri, and F. Martinelli. Classification of security properties - part ii: Network security. In *FOSAD*, pages 139–185, 2002. 102, 119

[111] A. Frier, P. Karlton, and P. Kocher. *The SSL 3.0 Protocol*. Netscape Communications Corp., Nov. 1996. 2, 99

[112] M. Garey and D. Johnson. *Computers and Intractability*. W. H. Freeman and Co., San Francisco, CA, 1979. 43

[113] O. Goldreich. *Foundations of Cryptography - Volume I (Basic Tools)*. Cambridge University Press, 2001. 3

[114] O. Goldreich. *Foundations of Cryptography - Volume II (Basic Applications)*. Cambridge University Press, 2004. 3

[115] S. Goldwasser and S. Micali. Probabilistic encryption. *Journal of Computer and System Science*, 28:270–299, 1984. DOI: 10.1016/0022-0000(84)90070-9 121

[116] L. Gong, R. Needham, and R. Yahalom. Reasoning About Belief in Cryptographic Protocols. In D. Cooper and T. Lunt, editors, *Proceedings 1990 IEEE Symposium on Research in Security and Privacy*, pages 234–248. IEEE Computer Society, 1990. DOI: 10.1109/RISP.1990.63854 99, 102, 104

[117] L. Gong and P. Syverson. Fail-stop protocols: An approach to designing secure protocols. *Dependable Computing for Critical Applications*, 5:79–100, 1998. 3, 120

[118] D. Gopan. *Numeric program analysis techniques with applications to array analysis and library summarization*. PhD thesis, Comp. Sci. Dept., Univ. of Wisconsin, Madison, WI, Aug. 2007. Tech. Rep. 1602. xi, 29

[119] S. Graf and H. Saïdi. Construction of abstract state graphs with PVS. In *Computer Aided Verif.*, volume 1254 of *Lec. Notes in Comp. Sci.*, pages 72–83, 1997. DOI: 10.1007/3-540-63166-6_10 12

[120] G. S. Graham and P. J. Denning. Protection — principles and practice. In *Proceedings of the AFIPS Spring Joint Computer Conference*, volume 40, pages 417–429. AFIPS Press, May 16–18 1972. DOI: 10.1145/1478873.1478928 69

[121] T. X. R. Group. The XSB programming system. http://xsb.sourceforge.net/. 33

[122] S. Gulwani and G. Necula. Precise interprocedural analysis using random interpretation. In *Princ. of Prog. Lang.*, 2005. DOI: 10.1145/1040305.1040332 23

[123] D. Harel, D. Kozen, and J. Tiuryn. *Dynamic Logic*. Foundations of Computing. MIT Press, 2000. 103

[124] M. A. Harrison and W. L. Ruzzo. Monotonic protection systems. In R. A. DeMillo, D. P. Dobkin, A. K. Jones, and R. J. Lipton, editors, *Foundations of Secure Computation*, pages 461–471. Academic Press, Inc., 1978. 67

[125] M. A. Harrison, W. L. Ruzzo, and J. D. Ullman. Protection in operating systems. *Communications of the ACM*, 19(8):461–471, Aug. 1976. DOI: 10.1145/360303.360333 63, 64, 66, 67

[126] W. Harrison. Compiler analysis of the value ranges for variables. *Trans. on Softw. Eng.*, 3(3):243–250, 1977. DOI: 10.1109/TSE.1977.231133 8

[127] C. He and J. C. Mitchell. Security analysis and improvements for IEEE 802.11i. In *Proceedings of the Network and Distributed System Security Symposium, NDSS 2005*. The Internet Society, 2005. 3, 99, 118

[128] C. He, M. Sundararajan, A. Datta, A. Derek, and J. C. Mitchell. A modular correctness proof of IEEE 802.11i and TLS. In *CCS '05: Proceedings of the 12th ACM conference on Computer and communications security*, pages 2–15, 2005. DOI: 10.1145/1102120.1102124 99, 121

[129] M. Hecht. *Flow Analysis of Computer Programs*. Elsevier Science Ltd., 1977. 6

[130] N. Heintze and J. D. Tygar. A model for secure protocols and their composition. *IEEE Transactions on Software Engineering*, 22(1):16–30, January 1996. DOI: 10.1109/32.481514 3, 120

[131] J. Herzog. *Computational Soundness for Standard Assumptions of Formal Cryptography*. PhD thesis, MIT, 2004. 122

[132] C. A. R. Hoare. An axiomatic basis for computer programming. *Communications of the ACM*, 12(10):576–580, 1969. DOI: 10.1145/363235.363259 103

[133] C. A. R. Hoare. Communicating sequential processes. *Commun. ACM*, 21(8):666–677, 1978. DOI: 10.1145/359576.359585 103, 105

[134] A. J. Hoffman and J. B. Kruskal. Integral boundary points of complex polyhedra. *Linear Inequalities and Related Systems*, pages 233–246, 1956. 43

[135] S. Horwitz, T. Reps, and D. Binkley. Interprocedural slicing using dependence graphs. *ACM Transactions on Programming Languages and Systems (TOPLAS)*, 12(1):26–60, January 1990. DOI: 10.1145/77606.77608 36

[136] S. Horwitz, T. Reps, M. Sagiv, and G. Rosay. Speeding up slicing. In *Proceedings of the Second ACM SIGSOFT Symposium on the Foundations of Software Engineering (FSE)*, pages 11–20, New York, December 1994. DOI: 10.1145/195274.195287 36

[137] J. Howell and D. Kotz. A formal semantics for SPKI. Tech. Rep. 2000-363, Department of Computer Science, Dartmouth College, Hanover, NH, Mar. 2000. 79, 84

[138] R. Impagliazzo and B. Kapron. Logics for reasoning about cryptographic constructions. In *Prof of 44th IEEE Symposium on Foundations of Computer Science (FOCS)*, pages 372–383, 2003. DOI: 10.1109/SFCS.2003.1238211 123

[139] R. Janvier, L. Mazare, and Y. Lakhnech. Completing the picture: Soundness of formal encryption in the presence of active adversaries. In *Proceedings of 14th European Symposium*

on Programming (ESOP'05), Lecture Notes in Computer Science, pages 172–185. Springer-Verlag, 2005. DOI: 10.1007/b107380 122

[140] S. Jha, N. Li, M. V. Tripunitara, Q. Wang, and W. H. Winsborough. Towards formal verification of role-based access control policies. *IEEE Trans. Dependable Sec. Comput.*, 5(4):242–255, 2008. DOI: 10.1109/TDSC.2007.70225 xi, 73

[141] S. Jha and T. Reps. Analysis of SPKI/SDSI certificates using model checking. In *Comp. Sec. Found. Workshop*, 2002. 91

[142] S. Jha and T. Reps. Model checking SPKI/SDSI. *J. Comp. Sec.*, 2004. 83, 91, 92

[143] T. Jim. SD3: A trust management system with certified evaluation. In *Proceedings of the 2001 IEEE Symposium on Security and Privacy*, pages 106–115. IEEE Computer Society Press, May 2001. DOI: 10.1109/SECPRI.2001.924291 2

[144] A. K. Jones, R. J. Lipton, and L. Snyder. A linear time algorithm for deciding security. In *17th Annual IEEE Symposium on Foundations of Computer Science (FOCS)*, pages 33–41, October 1976. DOI: 10.1109/SFCS.1976.1 63, 69

[145] J. Kam and J. Ullman. Monotone data flow analysis frameworks. *Acta Inf.*, 7(3):305–318, 1977. DOI: 10.1007/BF00290339 27

[146] M. Karr. Affine relationship among variables of a program. *Acta Inf.*, 6:133–151, 1976. DOI: 10.1007/BF00268497 23

[147] C. Kauffman. The Internet Key Exchange (IKEv2) protocol, 2005. RFC 4306. 3

[148] S. Kent and R. Atkinson. Security architecture for the internet protocol, 1998. RFC 2401. 2, 99

[149] G. Kildall. A unified approach to global program optimization. In *Princ. of Prog. Lang.*, pages 194–206, 1973. DOI: 10.1145/512927.512945 6, 13

[150] J. Knoop and B. Steffen. The interprocedural coincidence theorem. In *Comp. Construct.*, pages 125–140, 1992. DOI: 10.1007/3-540-55984-1_13 11, 12, 15, 20, 27

[151] J. Kohl and B. Neuman. The Kerberos network authentication service (version 5). IETF RFC 1510, September 1993. 2, 3, 99

[152] A. Lal, T. Reps, and G. Balakrishnan. Extended weighted pushdown systems. In *Computer Aided Verif.*, 2005. 21, 27

[153] B. W. Lampson. Protection. In *Proceedings of the 5th Princeton Conference on Information Sciences and Systems*, 1971. Reprinted in ACM Operating Systems Review, 8(1):18-24, Jan 1974. DOI: 10.1145/775265.775268 63

[154] K. Landfield. WU-FTPD resource center; personal communication. May 2003. 54

[155] W. Landi and B. Ryder. Pointer induced aliasing: A problem classification. In *Princ. of Prog. Lang.*, pages 93–103, Jan. 1991. DOI: 10.1145/99583.99599 27

[156] D. Larochelle and D. Evans. Statically detecting likely buffer overflow vulnerabilities. In *Proceedings of the 10^h USENIX Security Symposium*, August 2001. 35, 37, 55, 60

[157] P. Laud. Secrecy types for a simulatable cryptographic library. In *ACM Conference on Computer and Communications Security*, pages 26–35, 2005. DOI: 10.1145/1102120.1102126 122

[158] N. Li, B. N. Grosof, and J. Feigenbaum. Delegation Logic: A logic-based approach to distributed authorization. *ACM Transaction on Information and System Security*, 6(1):128–171, Feb. 2003. Preliminary versions appeared in *Proc. 1998 CSFW* and *Proc. 2000 IEEE Symposium on Security and Privacy*. DOI: 10.1145/605434.605438 2

[159] N. Li and J. Mitchell. Understanding SPKI/SDSI using first-order logic. In *Comp. Sec. Found. Workshop*, 2003. DOI: 10.1007/s10207-005-0073-0 91

[160] N. Li and J. C. Mitchell. RT: A role-based trust-management framework. In *The Third DARPA Information Survivability Conference and Exposition (DISCEX III)*. IEEE Computer Society Press, Apr. 2003. 2

[161] N. Li, J. C. Mitchell, and W. H. Winsborough. Design of a role-based trust management framework. In *Proceedings of the 2002 IEEE Symposium on Security and Privacy*, pages 114–130. IEEE Computer Society Press, May 2002. 92

[162] N. Li, J. C. Mitchell, and W. H. Winsborough. Beyond proof-of-compliance: Security analysis in trust management. *Journal of the ACM*, 52(3):474–514, May 2005. Preliminary version appeared in *Proc. 2003 IEEE Symposium on Security and Privacy*. DOI: 10.1145/1066100.1066103 xi, 63, 97, 98

[163] N. Li and M. V. Tripunitara. Security analysis in role-based access control. In *Proceedings of the Ninth ACM Symposium on Access Control Models and Technologies (SACMAT 2004)*, pages 126–135, June 2004. DOI: 10.1145/501983.502005 63, 73

[164] N. Li, W. H. Winsborough, and J. C. Mitchell. Distributed credential chain discovery in trust management. *Journal of Computer Security*, 11(1):35–86, Feb. 2003. Extended abstract appeared in *Proc. 2001 ACM CCS*. 93

[165] P. D. Lincoln, J. C. Mitchell, M. Mitchell, and A. Scedrov. Probabilistic polynomial-time equivalence and security protocols. In *Formal Methods World Congress, vol. I*, number 1708 in Lecture Notes in Computer Science, pages 776–793. Springer-Verlag, 1999. 4, 99, 101, 102, 123

[166] R. J. Lipton and L. Snyder. A linear time algorithm for deciding subject security. *Journal of the ACM*, 24(3):455–464, 1977. DOI: 10.1145/322017.322025 63, 69

[167] J. W. Lloyd. *Foundations of Logic Programming, Second Edition*. Springer, 1987. 32

[168] G. Lowe. Breaking and fixing the Needham-Schroeder public-key protocol using CSP and FDR. In *2nd International Workshop on Tools and Algorithms for the Construction and Analysis of Systems*. Springer-Verlag, 1996. DOI: 10.1007/3-540-61042-1_43 118

[169] G. Lowe. Some new attacks upon security protocols. In *Proceedings of 9th IEEE Computer Security Foundations Workshop*, pages 162–169. IEEE, 1996. 3, 99, 101, 102

[170] N. Lynch. I/O automata models and proofs for shared-key communication systems. In *Proceedings of 12th IEEE Computer Security Foundations Workshop*, pages 14–29. IEEE, 1999. DOI: 10.1109/CSFW.1999.779759 3, 120

[171] Z. Manna and A. Pnueli. *Temporal Verification of Reactive Systems: Safety*. Springer-Verlag, 1995. 113

[172] H. Mantel. On the Composition of Secure Systems. In *Proceedings of the IEEE Symposium on Security and Privacy*, pages 88–101, Oakland, CA, USA, May 12–15 2002. IEEE Computer Society. 121

[173] D. McCullough. Noninterference and the composability of security properties. In *Proceedings of the IEEE Symposium on Security and Privacy*, pages 177–186, Oakland, CA, USA, May 1988. IEEE Computer Society. DOI: 10.1109/SECPRI.1988.8110 121

[174] D. McCullough. A hookup theorem for multilevel security. *IEEE Transactions on Software Engineering*, 16(6):563–568, 1990. DOI: 10.1109/32.55085 121

[175] J. McLean. Security models and information flow. In *Proceedings of the IEEE Symposium on Security and Privacy*, Oakland, CA, USA, May 1990. IEEE Computer Society. 121

[176] J. McLean. A general theory of composition for a class of "possibilistic" properties. *IEEE Transactions on Software Engineering*, 22(1):53–67, 1996. DOI: 10.1109/32.481534 121

[177] C. Meadows. Applying formal methods to the analysis of a key management protocol. *Journal of Computer Security*, 1(1):5–36, 1992. 119

[178] C. Meadows. The NRL protocol analyzer: An overview. *Journal of Logic Programming*, 26(2):113–131, 1996. DOI: 10.1016/0743-1066(95)00095-X 99, 101, 102, 119

[179] C. Meadows. Analysis of the Internet Key Exchange protocol using the NRL protocol analyzer. In *Proceedings of the IEEE Symposium on Security and Privacy*. IEEE, 1998. 3, 99, 119, 120

[180] C. Meadows. Open issues in formal methods for cryptographic protocol analysis. In *Proceedings of DISCEX 2000*, pages 237–250. IEEE, 2000. DOI: 10.1109/DISCEX.2000.824984 120

[181] C. Meadows and D. Pavlovic. Deriving, attacking and defending the GDOI protocol. In *Computer Security - ESORICS 2004, 9th European Symposium on Research Computer Security, Proceedings*, volume 3193 of *Lecture Notes in Computer Science*, pages 53–72. Springer, 2004. 3, 99

[182] C. Meadows, P. F. Syverson, and I. Cervesato. Formal specification and analysis of the group domain of interpretation protocol using npatrl and the nrl protocol analyzer. *Journal of Computer Security*, 12(6):893–931, 2004. 119

[183] D. Micciancio and B. Warinschi. Completeness theorems for the Abadi-Rogaway logic of encrypted expressions. *Journal of Computer Security*, 12(1):99–129, 2004. Preliminary version in WITS 2002. 4, 122

[184] D. Micciancio and B. Warinschi. Soundness of formal encryption in the presence of active adversaries. In *Proceedings of TCC 2004*, volume 2951 of *Lecture Notes in Computer Science*, pages 133–151. Springer-Verlag, 2004. DOI: 10.1007/b95566 122

[185] J. K. Millen and V. Shmatikov. Constraint solving for bounded-process cryptographic protocol analysis. In *ACM Conference on Computer and Communications Security*, pages 166–175, 2001. DOI: 10.1145/501983.502007 118

[186] R. Milner. *A Calculus of Communicating Systems*. Springer-Verlag, 1982. 103, 105

[187] R. Milner, J. Parrow, and D. Walker. A calculus of mobile processes, i. *Inf. Comput.*, 100(1):1–40, 1992. DOI: 10.1016/0890-5401(92)90008-4 119

[188] R. Milner, J. Parrow, and D. Walker. A calculus of mobile processes, ii. *Inf. Comput.*, 100(1):41–77, 1992. DOI: 10.1016/0890-5401(92)90009-5 119

[189] N. H. Minsky. Selective and locally controlled transport of privileges. *ACM Transactions on Programming Languages and Systems*, 6(4):573–602, Oct. 1984. DOI: 10.1145/1780.1786 63

[190] J. Misra and K. M. Chandy. Proofs of networks of processes. *IEEE Transactions on Software Engineering*, 7(4):417–426, 1981. DOI: 10.1109/TSE.1981.230844 121

[191] J. C. Mitchell, M. Mitchell, and U. Stern. Automated analysis of cryptographic protocols using mur-phi. In *IEEE Symposium on Security and Privacy*, pages 141–151, 1997. DOI: 10.1109/SECPRI.1997.601329 99, 101, 102, 118

[192] J. C. Mitchell, A. Ramanathan, A. Scedrov, and V. Teague. A probabilistic polynomial-time calculus for analysis of cryptographic protocols (preliminary report). *Electr. Notes Theor. Comput. Sci.*, 45, 2001. DOI: 10.1016/S1571-0661(04)80968-X 4, 123

[193] J. C. Mitchell, A. Ramanathan, A. Scedrov, and V. Teague. A probabilistic polynomial-time process calculus for the analysis of cryptographic protocols. *Theor. Comput. Sci.*, 353(1-3):118–164, 2006. DOI: 10.1016/j.tcs.2005.10.044 4, 123

[194] J. C. Mitchell, V. Shmatikov, and U. Stern. Finite-state analysis of ssl 3.0. In *Proceedings of the Seventh USENIX Security Symposium*, pages 201–216, 1998. 118

[195] D. Moore, C. Shannon, and J. Brown. Code-red: a case study on the spread and victims of an internet worm. In *Proceedings ot the Internet Measurement Workshop 2002, Marseille, France*, November 6-8 2002. DOI: 10.1145/637201.637244 1

[196] R. Motwani, R. Panigrahy, V. A. Saraswat, and S. Ventkatasubramanian. On the decidability of accessibility problems (extended abstract). In *Proceedings of the Thirty-Second Annual ACM Symposium on Theory of Computing*, pages 306–315. ACM Press, May 2000. DOI: 10.1145/335305.335341 63

[197] S. Muchnick and N. Jones, editors. *Program Flow Analysis: Theory and Applications*. Prentice-Hall, Englewood Cliffs, NJ, 1981. 6

[198] M. Müller-Olm and H. Seidl. Precise interprocedural analysis through linear algebra. In *Princ. of Prog. Lang.*, 2004. DOI: 10.1145/964001.964029 19, 23, 24

[199] M. Müller-Olm and H. Seidl. Analysis of modular arithmetic. In *European Symp. on Programming*, 2005. DOI: 10.1145/1275497.1275504 19, 23, 25

[200] G. C. Necula, S. McPeak, and W. Weimer. CCured: type-safe retrofitting of legacy code. In *ACM SIGPLAN-SIGACT Conference on the Principles of Programming Languages (POPL)*, January 2002. DOI: 10.1145/565816.503286 35, 61

[201] R. Needham and M. Schroeder. Using encryption for authentication in large networks of computers. *Communications of the ACM*, 21(12):993–999, 1978. DOI: 10.1145/359657.359659 3, 101, 108, 121

[202] F. Nielson, H. Nielson, and C. Hankin. *Principles of Program Analysis*. Springer-Verlag, 1999. 6, 7

[203] S. Oh and R. S. Sandhu. A model for role admininstration using organization structure. In *Proceedings of the Seventh ACM Symposium on Access Control Models and Technologies (SACMAT 2002)*, June 2002. DOI: 10.1145/507711.507737 69

[204] P. on bugtraq mailing list. Technical analysis of the remote sendmail vulnerability. 4th March 2003. www.securityfocus.com/archive/1/313757 56

[205] C. Optimizer. *ILOG CPLEX Division*. 889 Alder Avenue, Incline Village, Nevada. www.cplex.com/ 42

[206] X. Ou, S. Govindavajhala, and A. W. Appel. Mulval: A logic-based network security analyzer. In *14th USENIX Security Symposium*, Baltimore, Maryland, August 2005. 2

[207] L. Paulson. Mechanized proofs for a recursive authentication protocol. In *Proceedings of 10th IEEE Computer Security Foundations Workshop*, pages 84–95, 1997. DOI: 10.1109/CSFW.1997.596790 99, 101, 102

[208] L. Paulson. Proving properties of security protocols by induction. In *Proceedings of 10th IEEE Computer Security Foundations Workshop*, pages 70–83, 1997. DOI: 10.1109/CSFW.1997.596788 99, 101, 102, 104, 118

[209] L. C. Paulson. Inductive analysis of the internet protocol tls. *ACM Trans. Inf. Syst. Secur.*, 2(3):332–351, 1999. DOI: 10.1145/322510.322530 99, 119

[210] B. Pfitzmann and M. Waidner. A model for asynchronous reactive systems and its application to secure message transmission. In *IEEE Symposium on Security and Privacy*, pages 184–, 2001. DOI: 10.1109/SECPRI.2001.924298 121

[211] M. Prasad and T.-C. Chiueh. A binary rewriting defense against stack based buffer overflow attacks. In *Proceedings of the USENIX'03 Annual Technical Conference*, June 2003. 35, 61

[212] A. Ramanathan, J. C. Mitchell, A. Scedrov, and V. Teague. Probabilistic bisimulation and equivalence for security analysis of network protocols. In *Foundations of Software Science and Computation Structures, 7th International Conference, FOSSACS 2004, Proceedings*, volume 2987 of *Lecture Notes in Computer Science*, pages 468–483. Springer-Verlag, 2004. DOI: 10.1007/b95995 4, 99, 101, 102, 123

[213] T. Reps, G. Balakrishnan, and J. Lim. Intermediate-representation recovery from low-level code. In *Part. Eval. and Semantics-Based Prog. Manip.*, 2006. DOI: 10.1145/1111542.1111560 10

[214] T. Reps, S. Horwitz, and M. Sagiv. Precise interprocedural dataflow analysis via graph reachability. In *Princ. of Prog. Lang.*, pages 49–61, 1995. DOI: 10.1145/199448.199462 11, 19, 20

[215] T. Reps, A. Lal, and N. Kidd. Program analysis using weighted pushdown systems. In *Found. of Software Tech. and Theoretical Comp. Sci.*, 2007. DOI: 10.1007/978-3-540-77050-3_4 xi, 13, 27

[216] T. Reps and G. Rosay. Precise interprocedural chopping. In *Proceedings of the Third ACM SIGSOFT Symposium on the Foundations of Software Engineering (FSE)*, volume 20, pages 41–52, October 1995. DOI: 10.1145/222124.222138 37

[217] T. Reps, S. Schwoon, and S. Jha. Weighted pushdown systems and their application to interprocedural dataflow analysis. In *Static Analysis Symp.*, pages 189–213, 2003. DOI: 10.1007/3-540-44898-5_11 12, 91

[218] T. Reps, S. Schwoon, S. Jha, and D. Melski. Weighted pushdown systems and their application to interprocedural dataflow analysis. *Sci. of Comp. Prog.*, 58(1–2):206–263, Oct. 2005. DOI: 10.1016/j.scico.2005.02.009 12, 13, 19, 22, 23, 25, 26

[219] A. Roy, A. Datta, A. Derek, and J. C. Mitchell. Inductive proofs of computational secrecy. To Appear. 4, 122

[220] A. Roy, A. Datta, A. Derek, and J. C. Mitchell. Inductive trace properties for computational security. In *ACM SIGPLAN and IFIP WG 1.7 7th Workshop on Issues in the Theory of Security*, 2007. 4, 122, 123

[221] A. Roy, A. Datta, A. Derek, J. C. Mitchell, and J.-P. Seifert. Secrecy analysis in protocol composition logic., 2006. to appear in Proceedings of 11th Annual Asian Computing Science Conference, December 2006. DOI: 10.1007/978-3-540-77505-8_15 102, 121

[222] R. Rugina and M. C. Rinard. Symbolic bounds analysis of pointers, array indices and accessed memory regions. In *ACM SIGPLAN Conference on Programming Language Design and Implementation (PLDI)*, 2000. DOI: 10.1145/349299.349325 35, 46, 61

[223] M. Rusinowitch and M. Turuani. Protocol insecurity with finite number of sessions is np-complete. In *CSFW*, pages 174–, 2001. DOI: 10.1109/CSFW.2001.930145 101, 119

[224] P. Ryan, S. Schneider, M. Goldsmith, G. Lowe, and B. Roscoe. *Modelling and Analysis of Security Protocols*. Addison-Wesley, 2001. 99, 101, 102, 105, 118

[225] M. Sagiv, T. Reps, and S. Horwitz. Precise interprocedural dataflow analysis with applications to constant propagation. *Theor. Comp. Sci.*, 167:131–170, 1996. DOI: 10.1016/0304-3975(96)00072-2 12, 15, 19, 23

[226] R. S. Sandhu. The schematic protection model: Its definition and analysis for acyclic attenuating systems. *Journal of the ACM*, 35(2):404–432, 1988. DOI: 10.1145/42282.42286 63, 69

[227] R. S. Sandhu. Expressive power of the schematic protection model. *Journal of Computer Security*, 1(1):59–98, 1992. 63

[228] R. S. Sandhu. The typed access matrix model. In *Proceedings of the 1992 IEEE Symposium on Security and Privacy*, pages 122–136. IEEE Computer Society Press, May 1992. DOI: 10.1109/RISP.1992.213266 63, 69

[229] R. S. Sandhu. Undecidability of the safety problem for the schematic protection model with cyclic creates. *Journal of Computer and System Sciences*, 44(1):141–159, Feb. 1992. DOI: 10.1016/0022-0000(92)90008-7 63

[230] R. S. Sandhu and V. Bhamidipati. Role-based administration of user-role assignment: The URA97 model and its Oracle implementation. *Journal of Computer Security*, 7, 1999. 69, 70

[231] R. S. Sandhu, V. Bhamidipati, and Q. Munawer. The ARBAC97 model for role-based aministration of roles. *ACM Transactions on Information and Systems Security*, 2(1):105–135, Feb. 1999. DOI: 10.1145/300830.300839 69, 70, 71

[232] R. S. Sandhu, E. J. Coyne, H. L. Feinstein, and C. E. Youman. Role-based access control models. *IEEE Computer*, 29(2):38–47, February 1996. DOI: 10.1109/2.485845 69

[233] R. S. Sandhu and Q. Munawer. The ARBAC99 model for administration of roles. In *Proceedings of the 18th Annual Computer Security Applications Conference*, pages 229–238, Dec. 1999. DOI: 10.1109/CSAC.1999.816032 69

[234] A. Schaad, J. Moffett, and J. Jacob. The role-based access control system of a European bank: A case study and discussion. In *Proceedings of the Sixth ACM Symposium on Access Control Models and Technologies*, pages 3–9. ACM Press, 2001. DOI: 10.1145/373256.373257 69

[235] S. Schneider. Security properties and CSP. In *IEEE Symp. Security and Privacy*, 1996. DOI: 10.1109/SECPRI.1996.502680 99, 101, 102

[236] S. Schneider. Verifying authentication protocols with csp. *IEEE Transactions on Software Engineering*, pages 741–58, 1998. DOI: 10.1109/32.713329 104, 118

[237] A. Schrijver. *Theory of Linear and Integer Programming*. Wiley, N.Y., 1986. 42, 43

[238] S. Schwoon. *Model-Checking Pushdown Systems*. PhD thesis, Technical Univ. of Munich, Munich, Germany, July 2002. 12, 16, 19, 21

[239] S. Schwoon, S. Jha, T. Reps, and S. Stubblebine. On generalized authorization problems. In *Comp. Sec. Found. Workshop*, 2003. 12, 83, 84, 88, 89, 91

[240] M. Sharir and A. Pnueli. Two approaches to interprocedural data flow analysis. In S. Muchnick and N. Jones, editors, *Program Flow Analysis: Theory and Applications*, chapter 7, pages 189–234. Prentice-Hall, Englewood Cliffs, NJ, 1981. 11, 12, 15, 27

[241] V. Shoup. On formal models for secure key exchange (version 4). Technical Report RZ 3120, IBM Research, 1999. 122

[242] N. P. Smith. Stack smashing vulnerabilities in the UNIX operating system. 1997. 35

[243] J. A. Solworth and R. H. Sloan. A layered design of discretionary access controls with decidable safety properties. In *Proceedings of IEEE Symposium on Research in Security and Privacy*, May 2004. DOI: 10.1109/SECPRI.2004.1301315 63

[244] D. Song. Athena: a new efficient automatic checker for security protocol analysis. In *Proceedings of 12th IEEE Computer Security Foundations Workshop*, pages 192–202. IEEE, 1999. DOI: 10.1109/CSFW.1999.779773 99, 102, 118, 119

[245] M. Soshi. Safety analysis of the dynamic-typed access matrix model. In *Proceedings of the Sixth European Symposium on Research in Computer Security (ESORICS 2000)*, pages 106–121. Springer, Oct. 2000. DOI: 10.1007/10722599_7 63

[246] M. Soshi, M. Maekawa, and E. Okamoto. The dynamic-typed access matrix model and decidability of the safety problem. *IEICE Transactions on Fundamentals*, E87-A(1):190–203,

Jan. 2004. 63

[247] P. Syverson and P. van Oorschot. On unifying some cryptographic protocol logics. In *Proceedings of 7th IEEE Computer Security Foundations Workshop*, pages 14–29, 1994. DOI: 10.1109/RISP.1994.296595 99, 102

[248] P. Syverson and P. van Oorschot. On unifying some cryptographic protocol logics. In *Proceedings IEEE Symposium on Research in Security and Privacy*, pages 14–28, 1994. DOI: 10.1109/RISP.1994.296595 104

[249] F. J. Thayer, J. C. Herzog, and J. D. Guttman. Mixed strand spaces. In *Proceedings of 12th IEEE Computer Security Foundations Workshop*, pages 72–82. IEEE, 1999. DOI: 10.1109/CSFW.1999.779763 3, 120

[250] A. F. Veinott and G. B. Dantzig. Integer extreme points. *SIAM Review*, 10:371–372, 1968. DOI: 10.1137/1010063 43

[251] C. Verbrugge, P. Co, and L. Hendren. Generalized constant propagation: A study in C. In *Comp. Construct.*, volume 1060 of *Lec. Notes in Comp. Sci.*, pages 74–90, 1996. DOI: 10.1007/3-540-61053-7_54 8

[252] D. Wagner. *Static Analysis and Computer Security: New techniques for software assurance.* PhD thesis, University of California, Berkeley, December 2000. 35, 37, 39, 60

[253] D. Wagner, J. S. Foster, E. A. Brewer, and A. Aiken. A first step towards automated detection of buffer overrun vulnerabilities. In *Network and Distributed System Security (NDSS)*, February 2000. 35, 56, 60

[254] S. Weeks. Understanding trust management systems. In *Proc. IEEE Symp. on Sec. and Privacy*, Oakland, CA, 2001. 76

[255] T. Y. C. Woo and S. C. Lam. A semantic model for authentication protocols. In *Proceedings IEEE Symposium on Research in Security and Privacy*, 1993. 114

[256] S. J. Wright. *Primal-Dual Interior-Point Methods.* SIAM, Philadelphia, 1997. 42

[257] R. Wunderling. *Paralleler und Objektorientierter Simplex-Algorithmus.* PhD thesis, Konrad-Zuse-Zentrum fur Informationstechnik Berlin, TR 1996-09. www.zib.de/PaperWeb/abstracts/TR-96-09/ 42, 46

[258] R. Wunderling, A. Bley, T. Pfender, and T. Koch. Sequential object-oriented simplex class library (SoPlex). www.zib.de/Optimization/Software/Soplex/ 42, 46

[259] H. Xi and F. Pfenning. Eliminating array bounds checks through dependent types. In *ACM SIGPLAN Conference on Programming Language Design and Implementation (PLDI)*, June 1998. DOI: 10.1145/277650.277732 60

[260] S. Yong, S. Horwitz, and T. Reps. Pointer analysis for programs with structures and casting. In *ACM SIGPLAN Conference on Programming Language Design and Implementation (PLDI)*, May 1999. DOI: 10.1145/301631.301647 53

Authors' Biographies

ANUPAM DATTA

Anupam Datta is on the research faculty at Carnegie Mellon University. Dr. Datta's research interests are in trustworthy systems, privacy, and analysis of cryptographic protocols. He has served as General Chair of the 2008 IEEE Computer Security Foundations Symposium, Program Co-chair of the 2008 Formal and Computational Cryptography Workshop, and on the program committees of many computer security conferences including ACM CCS, IEEE S&P, and IEEE CSF. Dr. Datta has a PhD in Computer Science from Stanford University and a BTech from IIT Kharagpur.

SOMESH JHA

Somesh Jha received his B.Tech from Indian Institute of Technology, New Delhi in Electrical Engineering. He received his Ph.D. in Computer Science from Carnegie Mellon University in 1996. Currently, Somesh Jha is a Professor in the Computer Sciences Department at the University of Wisconsin (Madison), which he joined in 2000. His work focuses on analysis of security protocols, survivability analysis, intrusion detection, formal methods for security, and analyzing malicious code. Recently he has also worked on privacy-preserving protocols. Somesh Jha has published over 100 articles in highly-refereed conferences and prominent journals. He has won numerous best-paper awards. Somesh also received the NSF career award in 2005.

NINGHUI LI

Ninghui Li is an Associate Professor of Computer Science at Purdue University. He received a Bachelor's degree from the University of Science and Technology of China in 1993 and a Ph.D. in Computer Science from New York University in 2000. Before joining the faculty of Purdue in 2003, he was a Research Associate at Stanford University Computer Science Department for 3 years. Prof. Li's research interests are in computer and information security and privacy, with focuses on access control and data privacy. He has published over 90 referred papers, and has served on the Program Committees of more than 50 international conferences and workshops, including serving as the Program Chair of the 2008 ACM Symposium on Access Control Models and Technologies and the 2009 IFIP WG 11.11 International Conference on Trust Management (IFIPTM). He is on the editorial board of the VLDB Journal. His research is funded by the National Science Foundation, the Air Force Office of Scientific Research (AFOSR), the Office of Naval Research (ONR), and by IBM and Google. In 2005, he was awarded an NSF CAREER award.

DAVID MELSKI

David Melski has been the head of the research division of GrammaTech, Inc. since 2002. Under Melski's leadership, GrammaTech Research focuses on automatic analysis and transformation of software for the purposes of reverse engineering, protection of critical IP, assurance, and producibility. GrammaTech Research employs static analysis, dynamic analysis, and combinations of static and dynamic techniques. GrammaTech Research is a leader in the development of analysis techniques both for source code and machine code. Melski received his Ph.D. in Computer Sciences from the University of Wisconsin, where his research interests included static analysis, profiling, and profile-directed optimization. Melski's Ph.D. thesis presented a framework for developing interprocedural path-profiling techniques, and examined the use of path profiles for automatic program optimization.

THOMAS REPS

Thomas Reps is Professor of Computer Science in the Computer Sciences Department of the University of Wisconsin, which he joined in 1985. Reps is the author or co-author of four books and more than one hundred fifty papers describing his research. His research has concerned program-development environments, software-engineering tools, incremental graph algorithms, program-analysis algorithms, and computer security. Reps received his Ph.D. in Computer Science from Cornell University in 1982, and his Ph.D. dissertation won the 1983 ACM Doctoral Dissertation Award. Reps has been the recipient of an NSF Presidential Young Investigator Award, a Packard Fellowship, a Humboldt Research Award, and a Guggenheim Fellowship. He is also an ACM Fellow. Reps has held visiting positions at the Institut National de Recherche en Informatique et en Automatique (INRIA) in Rocquencourt, France; the University of Copenhagen, Denmark; the Consiglio Nazionale delle Ricerche (CNR) in Pisa, Italy; and the Université Paris Diderot–Paris 7.